Disasters and History

Disasters and History offers the first comprehensive historical overview of hazards and disasters. Drawing on a range of case studies, including the Black Death, the Lisbon earthquake of 1755, and the Fukushima disaster, the authors examine how societies dealt with shocks and hazards and their potentially disastrous outcomes. They reveal the ways in which the consequences and outcomes of these disasters varied widely not only between societies but also within the same societies according to social groups, ethnicity, and gender. They also demonstrate how studying past disasters, including earthquakes, droughts, floods, and epidemics, can provide a lens through which to understand the social, economic, and political functioning of past societies and reveal features of a society which may otherwise remain hidden from view. This title is also available as Open Access on Cambridge Core.

Bas van Bavel is Distinguished Professor of Transitions of Economy and Society at Utrecht University.

Daniel R. Curtis is Associate Professor in the School of History, Culture and Communication at the Erasmus University Rotterdam.

Jessica Dijkman is Assistant Professor in Economic and Social History at Utrecht University.

Matthew Hannaford is a lecturer in Human Geography at the University of Lincoln.

Maïka de Keyzer is a lecturer in Socio-Economic and Environmental Pre-Modern History at KU Leuven.

Eline van Onacker is a policy analyst at the Department of Work and Social Economy of the Flemish Government.

Tim Soens is Professor of Medieval and Environmental History at the University of Antwerp.

Disasters and History

The Vulnerability and Resilience of Past Societies

Bas van Bavel
Utrecht University, The Netherlands

Daniel R. Curtis
Erasmus University Rotterdam, The Netherlands

Jessica Dijkman
Utrecht University, The Netherlands

Matthew Hannaford
University of Lincoln, UK

Maïka de Keyzer
KU Leuven, Belgium

Eline van Onacker
Department of Work and Social Economy of the Flemish Government, Belgium

Tim Soens
University of Antwerp, Belgium

CAMBRIDGE
UNIVERSITY PRESS

University Printing House, Cambridge CB2 8BS, United Kingdom

One Liberty Plaza, 20th Floor, New York, NY 10006, USA

477 Williamstown Road, Port Melbourne, VIC 3207, Australia

314–321, 3rd Floor, Plot 3, Splendor Forum, Jasola District Centre, New Delhi – 110025, India

79 Anson Road, #06–04/06, Singapore 079906

Cambridge University Press is part of the University of Cambridge.

It furthers the University's mission by disseminating knowledge in the pursuit of education, learning, and research at the highest international levels of excellence.

www.cambridge.org
Information on this title: www.cambridge.org/9781108477178
DOI: 10.1017/9781108569743

© Bas van Bavel, Daniel R. Curtis, Jessica Dijkman, Matthew Hannaford, Maïka de Keyzer, Eline van Onacker and Tim Soens 2020

This publication is in copyright. Subject to statutory exception and to the provisions of relevant collective licensing agreements, no reproduction of any part may take place without the written permission of Cambridge University Press.

An online version of this work is published at doi.org/10.1017/9781108569743 under a Creative Commons Open Access license CC-BY-NC 4.0 which permits re-use, distribution and reproduction in any medium for non-commercial purposes providing appropriate credit to the original work is given and any changes made are indicated. To view a copy of this license visit https://creativecommons.org/licenses/by-nc/4.0

All versions of this work may contain content reproduced under license from third parties. Permission to reproduce this third-party content must be obtained from these third-parties directly.

When citing this work, please include a reference to the DOI 10.1017/9781108569743

This book has obtained the Open Access Book Grant from the NWO (grant no. 36.201.006), which was received by Daniel R. Curtis as a result of leading an NWO VIDI project (grant no. 016.Vidi.185.046).

First published 2020

A catalogue record for this publication is available from the British Library.

ISBN 978-1-108-47717-8 Hardback
ISBN 978-1-108-70211-9 Paperback

Cambridge University Press has no responsibility for the persistence or accuracy of URLs for external or third-party internet websites referred to in this publication and does not guarantee that any content on such websites is, or will remain, accurate or appropriate.

Contents

List of Figures	*page* vii
List of Tables	viii
Preface	ix

1 Introduction: Disasters and History — 1
  1.1 The Key Themes of the Book — 1
  1.2 Disaster Studies and Disaster History: Connected Fields? — 10
  1.3 Interpretative Frameworks in Historical Research — 15

2 Classifications and Concepts — 22
  2.1 A Taxonomy of Disasters — 22
  2.2 Scale and Scope of Disasters — 25
  2.3 Concepts — 29
    2.3.1 Disaster and Hazard — 29
    2.3.2 The Disaster Management Cycle — 31
    2.3.3 Vulnerability — 33
    2.3.4 Resilience — 35
    2.3.5 Adaptation, Transformation, and Transition — 37
    2.3.6 Risk — 39

3 History as a Laboratory: Materials and Methods — 43
  3.1 Historical Sources — 44
    3.1.1 Types of Historical Sources — 44
    3.1.2 Combining Historical Data with Sources from the Natural Sciences — 53
    3.1.3 History and the Digital Age: Opportunities and Pitfalls for Historical Disaster Research — 57
  3.2 Methodologies — 61
    3.2.1 Hazard and Disaster Reconstruction from Historical Sources — 61
    3.2.2 Vulnerability Assessment — 65
    3.2.3 Comparative Methodologies — 68

4 Disaster Preconditions and Pressures — 74
  4.1 Environmental and Climatic Pressures — 74
  4.2 Technological, Infrastructural, and Economic Preconditions — 77
    4.2.1 Technological and Infrastructural Preconditions and Pressures — 77
    4.2.2 Economic Pressures and Crises — 81

	4.3 Coordination Systems and Institutional Preconditions	85
	4.3.1 Coordination Systems: The Family, the Market, and the State	86
	4.3.2 Institutions for Collective Action and the Commons	91
	4.4 Social Pressures: Poverty, Inequality, and Social Distress	96
	4.5 Cultural Preconditions	100
5	**Disaster Responses**	**105**
	5.1 Top-Down and Bottom-Up Responses	105
	5.2 Experience, Memory, Knowledge, and Experts	110
	5.2.1 Memory and Learning from Experience	110
	5.2.2 The 'Rule of Experts'	113
	5.3 Constraints on Disaster Responses	116
	5.3.1 Inequalities in Power and Property	116
	5.3.2 Institutional Rigidity and Path Dependency	120
6	**Effects of Disasters**	**123**
	6.1 Short-Term Effects	123
	6.1.1 Victims, Selective Mortality, and Population Recovery	123
	6.1.2 Land Loss and Capital Destruction	132
	6.1.3 Economic Crisis	135
	6.1.4 Scapegoating, Blame, and Social Unrest	137
	6.2 Societal Collapse	141
	6.3 Long-Term Effects	145
	6.3.1 Disasters as a Force for Good? Economic Effects	145
	6.3.2 Long-Term Demographic Changes	148
	6.3.3 Reconstruction, Reform, and Societal Change	150
	6.3.4 Economic Redistribution	154
7	**Past and Present**	**159**
	7.1 Disaster History and/in the Anthropocene	159
	7.1.1 Climate Change	161
	7.1.2 Capitalism	164
	7.1.3 The Risk Society	167
	7.2 The Potential of History for Better Understanding Disasters	169
	7.2.1 The Historical Roots of Present-Day Disasters	171
	7.2.2 The Past as an Empirical Laboratory: Institutions and Social Context	173
	7.2.3 The Great Escape: Can History Teach Us How to Escape from Disaster?	175
	7.3 The Potential of Disasters for Historical Research	177
	7.3.1 Disasters as Historical Protagonists	178
	7.3.2 Disasters as Tests at the Extreme Margin	179
	7.4 Future Pathways	181
	7.5 A Final Word on Disaster Victims	185
	References	188
	Index	225

Figures

1.1 Diversity of outcomes through hazards, disasters, and adaptation	*page* 2
1.2 The basic Malthusian model	17
2.1 The disaster cycle	32
3.1 The growth of instrumental meteorological observation, 1850–2012	51
3.2 Photos of the Manchurian plague of 1911	52
3.3 Illustration of geographical gaps in digitized Biraben plague datasets	59
4.1 George Pinwell, 'Death's Dispensary'	80
5.1 Engraving showing the drastic measures imposed by the Dutch government in 1745 to contain the rinderpest	107
6.1 Dust Bowl farm in the Coldwater District, north of Dalhart, Texas, June 1938	133
6.2 Painting by Thomas Cole, *The Course of Empire – Destruction*	144
7.1 Illustration of the core concepts of the WGII AR5	162
7.2 So-called 'scenario trumpet' projecting possible disaster scenarios	170

Tables

2.1 Ranked list of natural disasters by death toll on Wikipedia *page* 28
3.1 Historical documentary evidence for reconstructing hazards and their impacts prior to instrumental recording 45

Preface

This book has emerged in the context of four key developments occurring over the past decade or so in the broad field of 'the history of disasters.' These are, first, the growing recognition that natural conditions and events likely had an important role to play in determining historical outcomes; second, the increasing use of non-documentary sources – including DNA, tree rings, and ice-cores – and the related quest for linking the natural and social sciences; third, the establishment of very large digital databases of information; and fourth, the gradual dominance of an explanatory framework for disasters that emphasizes resilience, and, more specifically, adaptation. One of the goals of this book is to trace these developments and how they have recently shaped our understanding of the interaction between disasters and history. More significantly, however, we also offer some critique. Indeed, as a team of historians co-writing with an emphasis on social history, we suggest through the course of the book that although interaction with geneticists, climatologists, statisticians, and so on has brought many new advantages to our study of disasters in the past, we also need to continue to pay heed to 'traditional' historical skills – in particular the assessment of new data with regard to source critique and contextualization of evidence. Perhaps even more explicitly, however, given the emphasis on social history, we also put forward a view that we should not forget the significance of social relations and of disparities in advantages, opportunities, and access to resources between social groups. Indeed, history shows that while economic, institutional, agricultural, and ecological systems often remained resilient in the face of significant hazards – many of which led to disaster – significant social groups within those very same systems were often vulnerable, and suffered many hardships. We should be careful not to push these dimensions out of view.

The authors offer thanks to the editor, Michael Watson, and the staff of Cambridge University Press, and the three anonymous referees for their suggestions and comments on the first proposal for this volume. Similarly, we thank Bin Wong (UCLA) and Eleonora Rohland

(Bielefeld) for their comments on the first draft, and Heli Huhtamaa (Bern/Heidelberg), Joris Roosen (Utrecht) and Franz Mauelshagen (IASS Potsdam/Essen) for their comments and the extensive discussion we had at a workshop in Utrecht. Bram Hilkens (Erasmus University Rotterdam) and Constant van der Putten (Utrecht University) helped greatly with the technical editing, while Constant also took care of the illustrations, and Eileen Power helped us by checking the language. We are grateful to all of the people involved in this collective effort.

1 Introduction: Disasters and History

1.1 The Key Themes of the Book

This monograph provides an overview of research into disasters from a historical perspective, making two new contributions. First, it introduces the field of 'disaster studies' to history, showing how we can use history to better understand how societies deal with shocks and hazards and their potentially disastrous outcomes. Despite growing recognition of the importance of historical depth by scholars investigating disasters, the temporal dimensions of disasters have been underexploited up to now. Moreover, the historical record sometimes enables us to make a long-term reconstruction of the social, economic, and cultural effects of hazards and shocks that is simply not possible in contemporary disaster studies material. We can therefore use 'the past' as a laboratory to test hypotheses of relevance to the present in a careful way. History lends itself to this end because of the opportunity it offers to identify distinct and divergent social and environmental patterns and trajectories. We can compare the drivers and constraints of societal responses with shocks spatially and chronologically, and therefore enrich our understanding of responses to stress today.

Second, we introduce historians to the topic of disasters and the field of disaster studies, and explicitly show the relevance of studying past disasters to better understand the social, economic, and political functioning of past societies. Disasters often reveal features of society which in normal situations remain hidden from the view of the historian, for example, the entrenched vulnerability of particular groups within society or the manifestation of uneven power relations. People sometimes behaved in different ways during periods of pressure when compared with 'normal' times. Studying disasters thus allows historians to bridge the gap between 'event' and 'structure.' In particular, we show, through the lens of history and disasters, how the past can be used to carry out systematic spatiotemporal comparison and to empirically test hypotheses developed in the social and natural sciences. In this sense, the book looks to enrich approaches in the contemporary study of disasters, but also approaches and methodologies employed in the discipline of history.

The shocks and hazards on which this book concentrates are biophysical ones, including seismic activity, droughts, high water tables, and epidemics. Political and economic crises, war, and other human-made shocks may figure in the text, not per se, but as factors sharpening the effects of natural hazards or interacting with them. The broad objective of the book is to show how history can be used to demonstrate how these biophysical shocks and hazards, sometimes leading to disasters, push societies in different directions – creating a diversity of possible social and economic outcomes. Further, in this book we aim to identify the patterns and mechanisms involved in producing these outcomes.

This diversity of outcome is produced in three phases (see Figure 1.1). First, we show that the nature of the initial shocks to societies was often very different – some killing people but leaving capital untouched (such as the Black Death), others destroying capital and infrastructure but

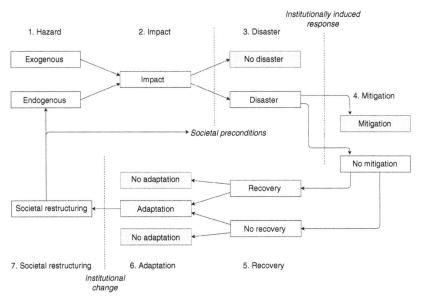

Figure 1.1 Diversity of outcomes through hazards, disasters, and adaptation, illustrating the framework used throughout this book. Drawn by Jasmin Palamar, Utrecht University.

1.1 The Key Themes of the Book

inflicting only modest casualties (such as most floods), and others destroying both at the same time (such as the most severe earthquakes, or epidemics occurring together with military activity). Put simply, the effects that different shocks had on production factors and the demands on infrastructure were not always the same. Even for the same type of shock there was great diversity: not all epidemics, for example, had universal features. Instead, they exhibited differences in epidemiologic characteristics on the grounds of severity, pervasiveness, longevity, seasonality, selectivity, and so on.

Second, we show that even when hazards and shocks had similar features, their outcomes often varied as a result of interacting with very different social, economic, political, and cultural settings, whereby dissimilar levels of pre-existing vulnerability acted as a 'filter' for the shock itself. Of course, famines and floods posed challenges very different from those of earthquakes or volcanic eruptions, but even societies interacting with the very same hazard were not always impacted in an identical way, nor did they always move in the same direction. Pre-existing vulnerability, a lack of preparedness, or ill-functioning institutions could turn a modest shock into a true disaster, while well-prepared and less vulnerable societies could withstand much bigger hazards and threats and even prevent them from turning into a disaster. Numerous kinds of institutional responses were formed at a variety of scales – some stronger on a supra-level through the state or long-distance markets, some stronger on a meso-level through collectives – such as guilds, community associations, organizations for the management of common-pool resources, insurance systems, and so on – and others stronger on a micro-level as households, families, and neighbors became the most dominant form of welfare, assistance. and protection. The arrangement and combination of these different scales of response helped create a diversity of outcomes.

Third, we show that even two different societies facing the same hazard with the same kinds of institutional responses still did not always produce the same post-hazard trajectories. As a major element to the book, we aim to demonstrate that although many historical societies shared a number of ways of dealing with the same shocks and hazards on the surface, for example in the form of poor relief institutions, flood management infrastructure, or commonly managed resources, the outcomes could still be quite different. In order to explain these differences, we look at the social actors behind the institutional responses themselves, showing that rather than being 'rational' responses offering the greatest amount of protection or welfare by way of institutional adaptation, these responses were the products of different social actors with goals not always equivalent to the 'common good.' Accordingly, we illustrate how shocks and hazards, and the disasters that ensued, could have very diverse consequences and outcomes, not only between societies, but also within the same societies, between social groups, and across wealth, ethnic, and gender lines, and we try to find patterns and mechanisms in order to better understand these diverse outcomes.

As well as focusing on diversity of outcomes, this book also addresses a number of key themes in disasters and history. An important one is that of anticipation, preparedness and memory. Hazards, and the disasters sometimes produced, are often not unexpected. Every region continuously had to cope with recurring sets of hazards and threats. Coastal zones, for instance, were constantly confronted by the threat of storm surges and floods, and therefore often formed a 'region of risk.'[1] This book addresses past societies' levels of preparedness, anticipation, and memory of hazards which emerge over the long term. Some societies were not only aware of the environmental threats but anticipated hazards and built a society that could cope with hazard reoccurrence.[2] Others, however, did not or could not anticipate and were therefore more vulnerable to unexpected environmental shocks. By looking into different cultural or social barriers, institutional flaws, and political constellations, we explain why certain societies were able to develop 'subcultures' of coping and introduced successful protection measures, while in others an effective response and level of preparedness was lacking. For example, some historians and historical geographers have suggested that 'cultural memory' of past hazards and disasters may have been important, helping initiate technological and infrastructure-based reorganization. Yet at the same time, memories of past events were sometimes manipulated by authorities – in the case of plague, for example, to act as a cautionary warning to citizens on how to behave.[3] Furthermore, mere knowledge of and experience with previous or repeat occurrences of hazards was not enough to stimulate adaptive practice in every case. That is to say, these responses were not always rational and effective reactions to protect the common good, but were sometimes used to promote the interests of select groups within society. Barriers could emerge to prevent successful adaptation regardless of knowledge and experience: in Southern Nyasaland (Malawi) after the 1949 famine, for example, peasants were largely prevented from using traditional ecological knowledge such as switching from maize to the more drought-resistant sorghum and from intensification of riverbank cultivation, by government policies that

---

[1] Bankoff, 'The "English Lowlands"'; Mauelshagen, 'Flood Disasters and Political Culture.'
[2] Bankoff, 'Cultures of Disaster.' [3] Carmichael, 'The Last Past Plague,' 159.

1.1 The Key Themes of the Book

prioritized the development of colonial cash crops at the expense of reducing sensitivity to drought.[4] Accordingly, coping mechanisms cannot be understood in isolation, but must be viewed in the wider context of societal organization and the associated institutional framework: that is, in relation to institutions not necessarily geared towards coping with hazards but serving very different goals.

Another key theme is that of disaster impacts. What are these diverse outcomes that we are able to reconstruct? Mortality is an obvious variable,[5] even though for some historical disasters precise data on the number and social profile of victims are lacking. Recent literature has shown how epidemics had different scales of severity and territorial reach not just between outbreaks, but also for the same outbreak across localities.[6] Likewise, it has been shown that neither periods of extreme weather nor periods of high prices in food products inevitably led to the same mortality effect.[7] Recent work on flooding has shown considerable regional differences in the disruption of livelihoods, the numbers of casualties produced, and the socio-economic profile of the victims.[8]

Mortality is not necessarily the optimal variable for all types of disasters, however. Some disasters did not cause high mortality but were disastrous on a different level. Floods or erosion could have devastating effects without killing a single person. For example, virtually nobody was killed 'directly' because of the sand storms during the American Dust Bowl in the 1930s. Nevertheless, this development has been labeled "one of the three worst ecological disasters in history,"[9] because of the loss of productive land that crippled the Midwestern economy indefinitely, with recent research showing significant medium- and long-term effects for human capital formation and for later-life health and income.[10] As a result, different variables could be applicable according to the type of disaster, such as capital destruction (in the form of land, labor, or financial capital), falling yields, or erosion of societal stability potentially leading to scapegoating or trauma. For historical disasters, capital destruction and changing yields or land use can be traced, as they left at least some marks in the written documentary sources, even if this is not always methodologically straightforward. A more complicated and less

---

[4] Vaughan, 'Famine Analysis.'   [5] Sen, 'Mortality.'
[6] Alfani, 'Plague in Seventeenth-Century Europe'; Curtis, 'Was Plague an Exclusively Urban Phenomenon?'
[7] Curtis & Dijkman, 'The Escape from Famine,' 235–236; Ó Gráda & Chevet, 'Famine and Market,' 714, 728.
[8] Rheinheimer, 'Mythos Sturmflut'; Elliott & Pais, 'Race, Class, and Hurricane Katrina'; Soens, 'Resilient Societies.'
[9] Quote from Borgström, *World Food Resources*, via Worster, *Dust Bowl*, 4.
[10] Arthi, "The Dust Was Long in Settling."

objective measure is whether a society experienced a societal shift out of its 'stability domain.' This is derived from the field of environmental studies, where a disaster is measured by the magnitude of disturbance that a system can absorb before it changes from one stability domain or ecosystem into another stable domain.[11] It could be that these forms of ecological resilience can be used for societies as well.[12] A hazard or shock could be labeled a disaster when a society is pushed out of its former stability domain into a new one. Returning to the American Dust Bowl, this is illustrated by the permanent shift that some regions made from being grain-producing regions towards cattle breeding in the places where the loss of productive soil was most severe.

Shocks and hazards could also stimulate different levels of social unrest. Although a clichéd view used to be that epidemics inevitably pushed all societies into disorder and disarray, perhaps even creating scapegoats, informed by a Foucauldian narrative of top-down repression, more recent work has tended to show oppositional tendencies too, as epidemics became forums for welfare reform, encouraged community cohesion, and provided a vehicle for those at the bottom of society to autonomously vent their frustrations and concerns towards those at the top.[13] Similarly, during periods of famine, it used to be seen as inevitable that the kinds of pressure generated thereby would break the bonds of society, leading to heightened levels of criminality such as thievery and violence, and yet other works in more recent years have shown how communal bonds of trust could continue and be strengthened even in some of the worst famines.[14]

Over the long term, we also find diversity in economic and demographic outcomes. Some of these were aggregate outcomes such as differences in overall extent and speed of population recovery after a disaster. Demographic development after the Black Death was not the same in the Low Countries as it was in England, Spain, or Egypt,[15] for example, and differed on a regional level too within these countries.[16] Differences were even seen on a city-by-city level: post-epidemic migration allowed some cities to recover within a matter of years, even exceeding previous populations, while others nearby saw complete contractions.[17] We look at how these differences were established through the various 'tools' for

[11] Gunderson, 'Ecological Resilience,' 427.   [12] Adger, 'Social Vulnerability.'
[13] Cohn, *Epidemics*; Curtis, 'Preserving the Ordinary.'
[14] Slavin, 'Market Failure'; Vanhaute & Lambrecht, 'Famine'; van Onacker, 'Social Vulnerability.'
[15] Van Bavel & van Zanden, 'The Jump-Start'; Malanima, *Pre-modern European Economy*; Borsch, *The Black Death*.
[16] Lewis, 'Disaster Recovery,' 792–793; van Bavel, 'People and Land,' 6–8.
[17] For post-Black Death Tuscany: Herlihy & Klapisch-Zuber, *Les Toscans et leurs familles*.

demographic recovery that societies had at their disposal – nuptiality, fertility, migration, welfare safety-nets, and grip over economic resources and opportunities.

Many of the long-term differences in economic outcomes we discuss, however, were redistributive. Hazards and shocks helped redistribute economic resources between social groups, therefore making societies more or less unequal. Indeed, scholars such as Branko Milanovic have suggested that "epidemics and wars alone can explain most of the swings in [pre-modern] inequality," while Walter Scheidel believes that the only time socio-economic inequalities leveled themselves out throughout history was during episodes of mass violence, destruction, or mortality.[18] By its very nature, this kind of redistribution creates diverse outcomes – giving to some while taking away from others within the same society – but the level of post-shock redistribution also differed from context to context. In some cases, a lasting equitable effect was seen – the Black Death and recurring plagues in Northern and Central Italy, for example, decreasing inequality for the next century or more, whereas yet other epidemics, such as the 1629–30 outbreak in Northern Italy, had only very brief and short-run equitable effects that quickly disappeared or were negated by certain institutions being employed to maintain the 'status quo'.[19] In contrast to the 'leveling hypothesis,' other shocks, however, pushed redistributive outcomes in the opposite direction, as certain groups 'instrumentalized' the hazard to their advantage – elites buffering destructive shocks to capital goods in times of floods or hoarding and speculating in times of famine.

Hazards and shocks also helped redistribute resources and opportunities between social groups in the same community. Disasters did not mean the same things for the very young or the very old as they did for working-age adults, or for 'native' inhabitants when compared with the experiences of 'recent' migrants, or for rural dwellers compared with those of the cities. Studies on contemporary disasters tend to emphasize, for example, how women are more susceptible than men to various negative outcomes from disasters – and yet historical work on the subject tends to find differential trajectories, dependent on shock and context. A long debate has ensued over the role of the Black Death in improving women's economic fortunes,[20] and yet it is clear that post-plague opportunities for women differed between regions, if we are to use indicators such as access to property, participation in economic roles outside the household, access to the marriage market, and access to care and

---

[18] Milanovic, *Global Inequality*, 62, 69; Scheidel, *The Great Leveler*, passim.
[19] See Section 5.3.2.   [20] Rigby, 'Gendering the Black Death.' See Section 5.3.2.

support.[21] The same holds for famines. While development economics tends to assert negative aspects for women emanating out of intra-household hierarchies dictating access to food or potential for abandonment, pre-modern famine history tends to show more diversity in outcomes – in some cases women experiencing a 'female mortality advantage' vis-à-vis men, or benefiting from their status as creditors with surplus capital or their control over food production, or reaping the benefits of poor relief systems structurally set up for women over men.

In discussing these themes, the book will not be comprehensive in its spatial coverage. Unfortunately, many parts of the globe can hardly be discussed because of the lack of sources or relevant studies. Things are changing for the better, however, with a growing number of studies appearing on the Caribbean, sub-Saharan Africa, and China, for instance. We hope this book will constitute an invitation to scholars working there. This will also enable us to see to what extent the concepts, definitions, and mechanisms presented in this book also apply to other parts of the globe. To be sure, in choosing cases to be discussed, this book does not attempt to be exhaustive. Rather, we feel that touching briefly on a very large number of cases could lead to superficiality and not be in line with one of the core arguments of the book, that is, that hazards and disasters can be understood and explained only by placing them in their social, economic, political, and cultural context. Therefore, we opted to somewhat limit the number of examples to be mentioned and also chose some well-investigated cases to return to throughout the book. Most notably, these cases are the Black Death of 1348 in Europe and the Middle East, the North Sea floods of the late-medieval and early-modern period, the Lisbon earthquake of 1755, the American Dust Bowl of the 1930s, the sub-Saharan Africa famines of the twentieth century, and the recent earthquake and tsunami in Japan leading to the Fukushima disaster. These cases relate to different types of hazards and offer different time scales (event versus long-run process), and they are relatively well documented and investigated, thus enabling us to highlight the various aspects of each disaster and its wider context.

This book does not limit itself to the modern period, as it takes a long-term perspective and tries to employ the whole of the historical record. Not all periods can be equally rigorously discussed, since there are limitations in terms of the sources and works available, but more attention than often is the case will be paid to the pre-industrial period. This reflects in part the academic background of the authors, but also the fact that this period knew a wide diversity of social, economic, cultural, and political

---

[21] See Section 6.1.1.

constellations, often found in close proximity to each other, and thus offers a very rich testing ground replete with opportunities to test ideas and hypotheses and find contrasting experiences that allow us to perform a comparative analysis.[22] Moreover, despite acknowledging that the Industrial Revolution brought momentous changes, we do not consider it to have formed a fundamental rift making the mechanisms at play in the pre-industrial period fundamentally different and therefore useless for understanding the challenges we are facing now; rather, many of the underlying mechanisms are essentially similar.

Linking up with this, the book explicitly inquires whether the present can be regarded as an 'Age of Disaster.' Higher vulnerability to disasters is often seen as one of the fundamental characteristics of the Anthropocene, the new geological epoch in which humanity influences the basic conditions of the planetary ecosystem directly.[23] In the Anthropocene, humans are co-producing nature – just as nature co-produces humans – but at the same time humans have to abandon the ambition to control nature, which had been one of the fundamental premises of 'modernity.'[24] The Enlightenment and the Industrial Revolution 'modernized' the disaster experience, as disasters were deemed to be both produced by the technological endeavors of men and controlled through technology. Since the late twentieth century, however, it has been argued by some scholars[25] that control is no longer possible, but risk reduction and adaptation are more crucial than ever, since the extreme events are there, we do not control them, and we cannot escape them.

While the scale and intensity of resource exploitation and technological transformation of the Earth undoubtedly accelerated hugely from the Industrial Revolution onwards, and became even more articulated after World War II, this book nevertheless questions whether pre-modern disasters were indeed very different from 'modern' or 'Anthropocentric' ones. At least they are not different in a fundamental way. First, modern advances did not free us from the risk of disaster. Seen from a long-term world-historical perspective, one could say that "every gain in precision in the coordination of human activity and every heightening of efficiency in production were matched by a new vulnerability to breakdown."[26] Second, and even more clearly, the social, political, cultural, and economic settings of societies, and the coordination systems they used – a main cause of the divergences in the effects of hazards and disasters

---

[22] Van Bavel & Curtis, 'Better Understanding Disasters.'
[23] Ebert, 'The Age of Catastrophe'.
[24] Chakrabarty, 'The Climate of History.' See Section 7.1.
[25] Beck, Lash & Wynne, *Risk Society*.
[26] McNeill, *The Global Condition*, 148; Mauelshagen, 'Defining Catastrophes.'

we highlight throughout the book – do not show a linear progression throughout history. Systems of competitive markets, and their dominance, for instance, can rise and decline, without an inevitable march forward to modernity.[27] This book thus seeks to replace a linear processional narrative on the modernization of disasters with a more analytical approach that focuses on continuities, disruption, and change in their production, interpretation, and (technological) control. In so doing, we question, for instance, the importance of technology in the production of and resilience to disasters long before the Industrial Revolution, and show the profound regional and social divergences in vulnerability and resilience characterizing 'traditional,' 'modern,' and 'Anthropocenic' societies.

The present experiences show that it is relevant, perhaps even more than ever, to use the historical record to increase our understanding of disasters. Societies all over the globe are confronted with rising water tables and ensuing floods, severe drought, or epidemics, including the COVID-19 pandemic. While the medium- and long-term consequences of these recent disasters will not be known to us for some time, we feel that the concepts, frameworks, and angles of analysis discussed in this book exploring the links between disasters and historical development, and the insights offered by historical analysis, may be fruitfully applied by other future scholars who will want to focus more explicitly on the fallout of present disasters. At the very least, the COVID-19 pandemic shows again how both over-emphasis on 'inevitable' mechanistic frameworks and processional narratives about progress and technology are unfounded and obscure the real effects on people, more particularly the different effects on different groups of people. This requires a better, deeper understanding of the causes of resilience and vulnerability of societies in the face of natural hazards, to which this book, with analysis based on the historical record, hopes to contribute.

1.2 Disaster Studies and Disaster History: Connected Fields?

The rise of interest in hazards and disasters as objects of scientific analysis is closely intertwined with the specifics of the Cold War period, when the US government and army became increasingly interested in how the American population would react if a (nuclear) attack occurred. In the 1950s the first research on how people reacted under extreme circumstances, where disasters such as fires, cyclones, and earthquakes were seen as a proxy for war, was conducted mainly by sociologists, looking at

---

[27] See for markets: van Bavel, *The Invisible Hand?*

the social effects of disasters. This continued in the 1960s when the Disaster Research Center (DRC) was founded by three disaster sociologists: Quarantelli, Dynes, and Haas.[28] At the same time, geographical interest in 'natural hazards' sparked, with a focus on building and land development.[29] Gilbert White's ground-breaking work on floodplain management famously stated: "Floods are an act of God, but flood losses are largely an act of man."[30]

Still, these early roots of disaster studies created a field that was mainly interested in the aftermath of disaster and focused on practical knowledge for disaster management. Disasters were seen as an event – as a short rupture of normalcy. In the 1990s, this was famously challenged by the work of Wisner and Blaikie, whose 'At Risk' argued that "although events such as hurricanes, floods, and earthquakes serve as triggers for disasters, disasters themselves originate in social conditions and processes that may be far removed from events themselves."[31] Anthropologists also developed a keen interest in disasters, often focusing on the bottom-up disaster experience of the 'subaltern,' adding new class, gender, and race perspectives to a field that had long been dominated by affluent white men.[32] These developments heavily nuanced the long-standing belief in the naturalness of disasters and the emphasis on the effects of disasters that had long dominated the field.

As climate change rose in importance on the political agenda, disaster studies increasingly concerned itself with anthropogenic climate change using the established concepts of vulnerability, resilience, and adaptation.[33] The UN call for the International Decade for Natural Disaster Reduction (IDNDR) during the 1990s and early 2000s played a part in this. Also very decisive was the 2004 'Reducing Disaster Risk' report by the UNDP. As climate change disproportionally impacted the developing world, the need for resilience and adaptation took center stage. This led to an increased interest in these topics in the social sciences, which trickled down to the field of history – albeit slowly.

Historians were relatively late to become involved.[34] Of course, individual cataclysms like the Black Death or the Lisbon Earthquake had always been of interest to economic, political, and cultural historians. In many ancient and pre-industrial cultures, the systematic recording and

---

[28] Quarantelli, 'Disaster Studies.'   [29] Tierney, 'From the Margins to the Mainstream?'
[30] White, 'Human Adjustment to Floods,' 2.
[31] Tierney, 'From the Margins to the Mainstream?,' 509.
[32] Oliver-Smith, 'Anthropological Research.'
[33] Steinberg, *Acts of God*; Groh, Kempe & Mauelshagen, *Naturkatastrophen*; Jakubowski-Tiessen & Lehmann, *Um Himmels Willen*.
[34] Steinberg, *Acts of God*; Groh, Kempe & Mauelshagen, *Naturkatastrophen*; Jakubowski-Tiessen & Lehmann, *Um Himmels Willen*.

interpretation of calamitous events even constituted an important part of the 'profession' of historians. In imperial China for instance, the official history of each dynasty included a subsection labeled 'the Five Phases' or 'Five Elements,' listing weather anomalies and disasters from floods via locusts to the appearance of dragons.[35] Dismissed as part of a 'positivist,' descriptive tradition, post-World War II historiography originally paid little attention to disastrous events, with some notable exceptions. In the 1970s mounting environmental concerns guided some historians to question more systematically the historical interaction between humans and nature, including the many moments when this interaction turned violent. For environmental history, as the new discipline was labeled in the United States, 'natural disasters' were an important subject from the very beginning. The ground-breaking work of Donald Worster on the American Dust Bowl, for instance, was one of the first studies to highlight the historical intertwinement of environment and economy, by showing how capitalist policies led to an aggressive exploitation of the Great Plains, paving the way for disaster.[36] In their attention to specific kinds of natural hazards and disasters, environmental historians often show 'national' preoccupations in line with the 'primary trauma' of their region: while environmental historians of the Low Countries automatically concentrated on water and floods, German historians were focusing on 'wood shortage' and the decline of forests, '*Waldsterben.*' For most non-European historians, the primary trauma was colonization, and both in Latin America and in India, a lot of historical literature on disasters aimed to reveal the 'colonial roots.'[37]

Historical climatology also paid ample attention to disasters induced by extreme weather conditions and/or climatic variability. However, while climate history today is very much concerned with the 'impact' of climate on society,[38] not all pioneers in the field initially were convinced of an intimate climate–society nexus. In the 1960s Emmanuel Le Roy Ladurie still considered climate history "history without human beings."[39] In his 1971 *Times of Feast, Times of Famine* he hence concluded that "in the long term the human consequences of climate seem to be slight, perhaps negligible, and certainly difficult to detect." A similar tone was later taken by Jan de Vries, who suggested that "short-term climatic crises stand in relation to economic history as bank robberies to the history of

---

[35] Brook, *The Troubled Empire*, 52. [36] Worster, *Dust Bowl*.
[37] Radkau, *Nature and Power*, 10. For the association with colonialism, see for instance Oliver-Smith, 'Peru's Five-Hundred-Year Earthquake.'
[38] Mauelshagen, 'Redefining Historical Climatology.'
[39] Le Roy Ladurie, *Histoire du climat*; Le Roy Ladurie, *Times of Feast*. See also Mauelshagen & Pfister, 'Vom Klima zur Gesellschaft.'

1.2 Disaster Studies and Disaster History

banking."[40] In contrast to Le Roy Ladurie, British climatologists Gordon Manley and Hubert Lamb made greater allowance for the influence of climate on human cultures, while being critical of the problems that characterized earlier determinist writings.[41] The work of Lamb and other researchers at the Climatic Research Unit (CRU) saw the formalization of historical climatology into a discipline that addressed "climate reconstruction, the identification and measurement of impact, and adaptation and perception."[42] During the 1980s, however, the emphasis of the research at the CRU would shift from climate impact to statistical climatology and climate modeling.

Only in the last couple of decades has historical climatology once again placed the climate–society nexus at the center of its work, with Christian Pfister – one of the few historians to contribute to both climate reconstruction and climate impact studies – identifying the vulnerability of past societies to climatic variation as a new focus for historical climatology in 2010.[43] During the past few years, climate history has boomed as never before. Periods of extreme 'climate stress' – pronounced stretches of extreme weather, or even shifts in the global climate system – are increasingly singled out as 'historical drivers' of major societal changes. Climate histories have been written for regions and societies outside of the traditional hotbed of central and western Europe, while increasing numbers of high-profile studies written by economic, social, and political historians have appeared on the environmental drivers of the decline of the Roman Empire, dynastic changes in Imperial China, or the spread of the Black Death in the fourteenth century.[44]

Apart from climate history, there is yet another historical subdiscipline which traditionally paid great attention to crisis and disaster. For socioeconomic historians, rising grain prices, food shortages, and famine have been a long-standing topic of interest.[45] The recurrence of 'subsistence crises' – as they were labeled by Meuvret – was considered an intrinsic feature of any pre-twentieth-century economy which only industrialization and the modernization of agricultural production could overcome.[46] Indeed, famine had always been high on the research agenda of economic historians, though it was a development economist, Amartya Sen, who

---

[40] De Vries, 'Measuring the Impact of Climate on History,' 603.
[41] Manley, *Climate and the British Scene*; Lamb, *Climate: Past, Present and Future*; Lamb, *Climate, History*.
[42] Ingram, Underhill & Wigley, 'Historical Climatology'; Wigley, Ingram & Farmer, *Climate and History*; Rotberg & Rabb, *Climate and History*.
[43] Pfister, 'The Vulnerability of Past Societies.'
[44] Harper, *The Fate of Rome*; Parker, *The Global Crisis*; Campbell, *The Great Transition*.
[45] To cite but one example: Beveridge, 'Wheat Prices and Rainfall' was published in 1922.
[46] Meuvret, 'Les oscillations des prix.'

forcefully demonstrated the potential of historical famines – in his case the 1943 Bengal Famine – to improve our understanding of present-day hunger. Sen's interpretation of famine as allocation or entitlement crises sparked a renewed interest in historical famines, lasting until today.[47] Epidemics – and especially the Black Death – and their economic consequences have also received a lot of attention, again often linked to the history of prices and wages. Whether the strong demographic contraction after the Black Death and the changing land–labor ratio led to a 'Golden Age of Labour' has been part of a historiographical debate ever since the nineteenth century (the term itself was coined in 1884), referring to the spike in wages and drop in prices observed by the first price and wage data collectors.[48] Debates on the same issue continue today.[49] The consequences of the Black Death not only inspired economic historians, but also attracted attention to cultural aspects, such as its impact on religion, the scapegoating of vulnerable groups, and the evolution of medical thinking.[50]

Moreover, we should keep in mind that the study of past disasters is not the fief of historians alone. In recent years archaeologists, anthropologists, geographers, and climate scientists have all enthusiastically embraced the potential of 'natural archives' – from sediments to ice cores – to reconstruct, date, and interpret the role of both extreme events and long-term changes in the rise and decline of communities, societies, and empires in the past. It has been argued, for instance, that the giant Laacher See volcanic eruption (in the Eifel region between Bonn and Koblenz) 13,000 years ago might explain the sudden disintegration of the homogeneous 'Federmesser' culture of hunter–gatherer communities in many parts of Northern and Western Europe.[51] In similar ways, a large volcanic eruption in March 536 AD is now thought to be responsible for disrupting climatic conditions, setting the scene for the outbreak of the Justinianic plague, ravaging the Byzantine Empire, and marking the beginning of a 'Late Antique Little Ice Age.'[52]

But perhaps the potential of disaster history might be situated not so much in formulating or rebutting this kind of grand narrative on the rise and fall of civilizations, but rather in revealing the causal mechanisms

---

[47] Sen, *Poverty and Famines*; Appleby, 'Grain Prices'; Ó Gráda & Chevet, 'Famine and Market'; Vanhaute & Lambrecht, 'Famine.'
[48] Rodgers, *Six Centuries*, 326.
[49] For an overview: Hatcher, 'Unreal Wages'; Hatcher & Bailey, *Modelling the Middle Ages*.
[50] Herlihy, *The Black Death*; Cohn, *The Black Death Transformed*; Cohn, *Cultures of Plague*; Cohn, *Epidemics*.
[51] Riede et al., 'A Laacher See-Eruption Supplement'; Riede, 'Towards a Science of Past Disasters.'
[52] Büntgen et al., 'Cooling and Societal Change.'

which explain why particular disasters did so much harm in some contexts and were countered in others. Over the past decade, this approach has been refined and tested by a group of historians in the Low Countries, including the authors of the present volume. For a wide array of hazards and disasters, from floods to epidemics and epizootics, from famine to sand drifts, the often highly contrasting regional experiences have been analyzed in depth. Such a comparative approach allowed us to question the relative impact of natural variability and exogenous shocks, but also of different coordination systems, state interventions, solidarity mechanisms, economic inequality, and so on. Understanding the changes in land-use which made a region vulnerable to devastating sand drifts[53], the marginalization processes which exposed certain regions or households to flood risk,[54] or the different recovery rates of cities after epidemics[55] turns disaster history into a useful empirical 'laboratory' to improve our knowledge of disasters in the past,[56] and, at the same time, sheds new light on the functioning of past societies.[57]

1.3 Interpretative Frameworks in Historical Research

Long before the history of disasters and its vocabulary of risk, vulnerability, resilience, and adaptation became fashionable, historians investigated major societal 'shocks' and developed interpretations on how to explain the origins and impact of these shocks, and regional variations in their frequency or intensity. From a cultural angle, for example, the inflexibility of cultural practices has been presented as one of the explanations for the failure of some societies to adapt to changing ecological circumstances.[58] Another line of thought has identified the nature of political regimes – democratic or dictatorial – and the accompanying 'inclusive' or 'extractive' institutions as the main determinants of success or failure.[59] While we acknowledge the importance of these approaches in cultural and political history, and will use them throughout the book, we will expand here on those developed within economic and social history, where perhaps the earliest attempts were made to arrive at overarching interpretations, which often hold direct relevance to the issues of vulnerability and resilience to hazards and shocks. In these fields, a number of interpretative schemes, theories, or models have emerged, aiming to explain not only economic

---

[53] De Keyzer & Bateman, 'Late Holocene Landscape Instability.'
[54] Soens, 'Resilient Societies.'
[55] Curtis, 'Was Plague an Exclusively Urban Phenomenon?'
[56] Van Bavel & Curtis, 'Better Understanding Disasters.'
[57] Curtis, van Bavel & Soens, 'History and the Social Sciences.'
[58] Krüger et al., *Cultures and Disasters*.   [59] Acemoglu & Robinson, *Why Nations Fail*.

growth and prosperity, but also the opposite: crisis and collapse. In particular, these approaches have been applied and tested on the Black Death, which may have killed half or more of the populations that it affected in Eurasia and parts of Africa in the middle of the fourteenth century. Over the past century, debates on the causes and consequences of the 'late-medieval crisis' and its possible connection to the Black Death have been structured along the lines of four major explanatory frameworks, which still influence, albeit often implicitly, the thinking about hazards and disasters in history more generally.[60] These frameworks center respectively around (i) population and resources, (ii) the social distribution of power and property, (iii) commercialization and markets, and (iv) institutions.

The first framework is often termed a 'Malthusian' approach, whereby hazards and shocks are the 'positive checks' that stem from the pressure between growing populations in a world with finite resources (Figure 1.2). These increased populations become faced with increasing food prices and rents, while labor at the same time becomes cheaper – thus real income declines. Either populations adapt to these conditions (by reducing the birth rate and hence population growth), forming a preventive check, or they face malnutrition and become vulnerable to epidemic diseases, which forms the positive check that adjusts populations to the available resources. For Sir Michael Postan, for example, the tension between population and resources was at the heart of the so-called late-medieval crisis. In famine analysis, this is translated into the tension between the availability of food and the population which needs to be fed. As long as the productivity of land remained inherently limited, the periodic recurrence of food shortage was seemingly inevitable. Examples of these limitations, with all the associated vulnerabilities, are found not only in thirteenth-century Europe. Late-Imperial China is also mentioned as a classic example of a technological–environmental 'lock-in,' where the perfection of available technologies and knowledge had pushed the productivity of the soil (and hence population density) far beyond what was 'sustainable,' necessitating ever higher investments to maintain the 'equilibrium' (or postpone the collapse). The inherent tension between population and resources made the region increasingly vulnerable to climatic variability, and led to hazards such as floods and harvest failures.[61]

That is not to say that every climate-centered analysis of disaster accords with a Malthusian framework. For medieval historians like

---

[60] Hatcher & Bailey, *Modelling the Middle Ages*, resumes the debate and its development. Literature on the different approaches framing the Black Death is extensive, see Campbell, *The Great Transition*, for the most recent synthesis.
[61] Elvin, *The Retreat of the Elephants*, esp. Chapter 1; Elvin, 'Three Thousand Years of Unsustainable Growth.'

1.3 Interpretative Frameworks

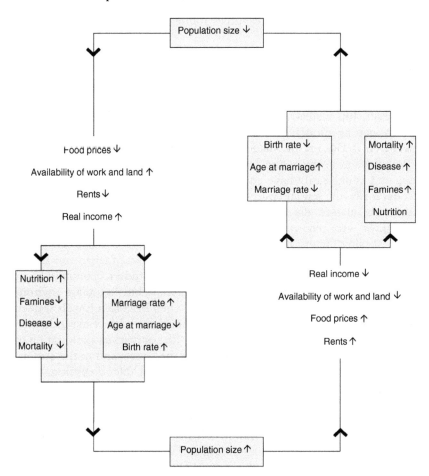

Figure 1.2 The basic Malthusian model, based on the model of John Hatcher and Mark Bailey. Hatcher and Bailey, *Modelling the Middle Ages*.

Bruce Campbell and David Herlihy, the thirteenth century "status quo of a maximum population subsisting with minimum living standards" could have continued almost indefinitely, had it not been brutally disrupted by the climatic and epidemic upheavals of the fourteenth century.[62] Put simply, climate- and epidemic-related shocks can also be seen as entirely

---

[62] Campbell, 'Nature as Historical Protagonist,' 287–288; quoting Herlihy, *The Black Death*, 38.

'exogenous' to the people–resources framework rather than having any causal relationship to it.

The main challenge to Malthusian interpretations of the late-medieval crisis, however, was offered by the Marxist analyses of class struggle and the allocation of the means of production between different groups in society. In this second approach, crises did not occur through a lack of resources, but instead because the political economy prevented social groups getting enough access to food or resources. As E. A. Kosminsky stated in 1956 on the late-medieval population crisis in England: "Probably, even given the level of productive forces then prevailing, England could easily have supported a much larger population, if the feudal lords, the feudal church and the feudal state had not sucked the labouring classes dry."[63] Depending on the historical context, the 'bad guys' were rent-extracting feudal lords or urban elite governments, profit-seeking factory owners, or 'enlightened' state officials seeking to rationalize economic production, even if we see cases of tenant-on-tenant extortion within the peasantry too.[64] In Marxist studies, economic crises are an inevitable consequence of elite extraction, but so too are environmental crises. For example, in the 1840s, when Friedrich Engels walked through the streets of Manchester – the 'shock city' of the First Industrial Revolution – he was appalled not only by "the barbarous exploitation of the workers," but also by the "foul air" in the streets and in the cotton- and flax-spinning mills where fibrous dust caused "blood-spitting, hard, noisy breathing, pains in the chest, coughs, sleeplessness," the lack of cleanliness and comfort in the houses, and the "narrow, coal-black foul-smelling" River Irk.[65] What Engels described was the unfolding of a 'slow disaster' – the gradual deterioration of living conditions which put people at risk of early death through chronic illnesses, and made them vulnerable to 'fast disasters,' which in the industrializing cities of the nineteenth century took the form of cholera and tuberculosis epidemics, toxic leaks, and mining catastrophes.[66] Some of these deteriorating urban conditions even predated the Industrial Revolution – becoming more of a problem through proletarianization and migration in the early-modern period.[67]

The rural counterpart of this approach is found in the long tradition in peasant history of investigating processes of land grabbing, expropriation, and the privatization of formerly common resources and the way they

[63] Kosminsky, *Studies in the Agrarian History of England*.
[64] For this nuanced view: Campbell, 'The Agrarian Problem.'
[65] Engels, *The Condition of the Working Class*; quotations by Clark & Foster, 'The Environmental Conditions of the Working Class.'
[66] Platt, *Shock Cities*. [67] Van Oosten, 'The Dutch Great Stink.'

1.3 Interpretative Frameworks

eroded the sustainability of the peasants' livelihoods. For Ramachandra Guha, the clash between the shifting forest cultivation – *jhum* – practiced by peasant communities in the Northeast of India and the commercial forestry advocated by the British Imperial Forestry department was not only a clash between fundamentally opposed ideas on how to use forest resources, but also a "struggle for existence" between villagers and the Forest Department, and between subsistence and the market. The progressive erosion of the *jhum* turned many villagers into landless laborers said to be more vulnerable than before.[68] Since Immanuel Wallerstein's work on economic world-systems, conflicting social relations are no longer situated solely between different classes within society, but also between regions at the core and periphery of the gradually emerging capitalist world-economy.[69] The shifting frontiers of the world-economy are often thought to have been particularly vulnerable to disasters, as the never-ending search for cheap labor and cheap natural resources often abruptly and radically transformed localized societies and environments, leading to over-exploitation of land and resources, and massive delocalization of people.[70] Recently, Marxist historians have developed an even more environmental approach by focusing on the 'metabolic rift,' a concept introduced by Karl Marx. In this literature, a capitalist mode of production, with a focus on relentless economic growth and strict division of labor, inevitably leads to environmental problems, that may result in true disasters. At the heart is a societal blueprint creating social vulnerabilities such as inequality, precariousness, weak entitlements, and monoculture, which exacerbate the effects of hazards and shocks.[71]

A third main approach to understanding crises and disasters, and their prevention or mitigation, is market dynamics – also labeled the Smithian or modernization approach. This framework focuses on economic growth through commercialization and markets, whereby expanding markets gave producers incentives to specialize, and the growing division of labor allowed economies of scale and productivity gains. These productivity gains in turn allowed living standards to grow in parallel with population, in contrast to Malthusian predictions, potentially postponing or even preventing crisis.[72] Furthermore, productivity growth offered

---

[68] Guha, *The Unquiet Woods*; Guha & Gadgil, 'State Forestry and Social Conflict.'
[69] Wallerstein, *The Modern World-System I*.
[70] Moore, 'The Capitalocene.' See Section 7.1.2.
[71] Foster, 'Marx's Theory of Metabolic Rift'; Moore, 'Environmental Crises and the Metabolic Rift.'
[72] For medieval Europe, see for instance Britnell, *The Commercialisation of English Society*. See also Section 6.1.3.

surpluses that could be reinvested to reduce pollution or develop protective technologies. Similar arguments have been extended to factor markets: clearly delimited and legally sanctioned private property rights, backed up by the enforcing powers of the state, have been judged by some to stimulate economic development and reduce exposure to disasters.

Of course, there is a reverse side to this kind of explanatory framework since many of these developments such as commercialization and market orientation also gave rise to knock-on developments such as social polarization, proletarianization, inequality, displacement, and so on – essential features guiding heightened vulnerability that have been discussed in the above-mentioned property-rights approach. Put simply, surpluses may have been produced, but in whose hands did they fall? Many of the investments in production often came hand-in-hand with restrictive regulations – forcing producers to cultivate certain types of crop or enter into inequitable credit agreements leading to debt bondage.[73] While markets could mitigate food shortages, they could also lead to speculation and hoarding – thus conversely making shocks such as harvest failures more severe.[74] Accordingly, whether market expansion, economic growth, and 'modern' clear property rights increased vulnerability to shocks, or whether they were the key behind an 'escape from disaster,' is a much debated subject which is explicitly dealt with elsewhere in this volume.[75]

In this context it is important to note that markets differ widely in their institutional organization, and this leads us to the fourth main approach to explaining crises and disasters, which focuses on the rules of human interaction: the institutions. Over the past decade, many innovative contributions to this issue, both in economics and in economic history, have been situated in the field of (New)-Institutional Economics. Institutional approaches often focus on the role of property rights and on the role of the state as third-party enforcer of clear and secure property rights, offering stability and stimuli for investment and potentially enhancing the resilience of societies to shocks and disasters[76] The state, however, may also be part of the problem instead of its solution: when rulers or governments extract part of the productive surplus for their own profit, or allow privileged groups in society to do so, they may constrain the potential for growth and increase vulnerability.[77] Also, even though institutional approaches often tend to highlight private property rights, alternative property regimes have also been considered. The recognition of the

[73] Van Bavel, *The Invisible Hand?*, 114–119. [74] Galloway, 'Basic Patterns,' 277.
[75] See Sections 7.2.2 and 7.2.3.
[76] See the important works by Douglass North in particular.
[77] Acemoglu & Robinson, *Why Nations Fail*, esp. Chapter 5.

importance of common property regimes in managing hazards and avoiding disasters has increased significantly over the past decades, not least thanks to the work of Elinor Ostrom, who argued that well-managed and well-delimited commons could well prevent a presumably inevitable Tragedy of the Commons.[78]

Institutional arrangements might be the key to understand why the same hazard turned into a full-blown disaster in one case, but not in others. The differential impact of the Black Death is once again a case in point. Comparing the opposing outcomes of the plague in Egypt and England, Stuart Borsch has argued that the causes of Egypt's economic decline, as opposed to England's recovery, are to be found in two contrasting systems of landholding. Whereas English landlords, holding their land as hereditary fiefs, were usually closely involved in the management of their estates and sufficiently interested in reviving their profitability in the changed post-Plague world to bargain with peasants and tenants, the short-term and non-hereditary landholding structure in Egypt stimulated a different attitude. Mamluk and *amir* landholders were not in direct contact with the peasants on their scattered estates but relied on an extensive bureaucracy supplemented by collective military expeditions in cases of social unrest. Those mechanisms, however, no longer functioned after the Plague as they had done earlier, since labor became scarce and intra-elite coherence crumbled. Vital irrigation systems were no longer maintained, agrarian productivity plummeted and rural depopulation set in.[79] The comparison between Egypt and England illustrates a general point this book wants to make: when analyzing and explaining how, and whether, hazards turn into disasters, we should not limit the analysis to institutional arrangements directly related to the governance of a particular hazard, but should include the whole arrangement of economy and society.

---

[78] Ostrom, *Governing the Commons;* De Moor, *The Dilemma of the Commoners.* See section 4.3.2.
[79] Borsch, *The Black Death in Egypt and England,* 26–27, 32–33, 40–41, 55–56.

2 Classifications and Concepts

2.1 A Taxonomy of Disasters

Given that there is a wide range of causes and consequences of disasters, it is unsurprising that there are also numerous forms of disaster classification. A classic categorization, focusing on the causes of disasters, is the distinction between what is natural and what is human-made. For more than one reason the validity of such a simple dichotomy is questionable. Indeed, scholarship on Hurricane Katrina has already claimed that "there is no such thing as a natural disaster."[1] Even though the initial shock was a natural event, the catastrophic outcome was ultimately the result of human intervention – or the lack of it. Put simply, without existing societal vulnerability, the chances of a hazard turning into a disaster are small.[2] Sometimes, the hazard or shock itself is also partly human-induced. For example, in the Limbe region in Cameroon, landslides are triggered by intense rainfall together with deforestation of steep slopes, soil excavation, and unregulated building activities.[3] The difficulty with this kind of classification system then is the blurred line between what might be considered 'exogenous' and 'endogenous,' with both features frequently present.

Another classic typology, also starting from causes, is a distinction based on the type of event that triggered the disaster, and is commonly used in contemporary disaster management. Leaving aside human conflict and industrial and transport-related accidents, three broad categories can be identified: disasters triggered by geological events (such as earthquakes, volcanic eruptions, and landslides), biological events (such as epidemics and epizootics) and meteorological–hydrological events (such as storms, floods, and droughts).[4] Sub-divisions and variations are

---

[1] As reads the title of an edited volume: Hartman & Squires (eds.), *There Is No Such Thing*.
[2] Blaikie *et al.*, *At Risk*, 45.   [3] Che *et al.*, 'Systematic Documentation.'
[4] Eshghi & Larson, 'Disasters.'

2.1 A Taxonomy of Disasters

possible. Famines, for instance, are often distinguished as a separate category – perhaps as a result of the complexities involved in explaining them.[5] There is, however, a tendency for crossovers and combinations: meteorological–hydrological events such as storms may lead to harvest failure and to famine, and they in turn may lead to biological events such as epidemics.

A different approach focuses on the time it takes for the triggering event to build up and the disaster to actually unfold, distinguishing between 'rapid onset' disasters, such as earthquakes and hurricanes, and 'slow onset' or insidious disasters, which include various types of environmental degradation such as desertification, sand drifts, and sea-level rise.[6] This categorization only partially overlaps with a typology based on triggering events. While many geological hazards – particularly earthquakes – are of the rapid onset type, meteorological–hydrological hazards can fall into either of the two categories: temporally, a hurricane is very different from rising sea levels. Biological hazards, moreover, fall somewhere in between: the unfolding of an epidemic can take months or years rather than minutes, but with differing stages of intensity. Rapid onset disasters, taking people by surprise, are more likely to lead to high levels of physical destruction, mortality, and displacement of survivors. The Lisbon earthquake of 1755, for instance, destroyed the city almost completely. Recent estimates arrive at a death toll of 20,000 to 30,000 (from a population of 160,000 to 200,000) with a similar number of survivors leaving the city.[7] Slow onset disasters are more insidious and therefore rarely cause immediate death; they are more likely to impact livelihoods in the long run. Over the long term, hazards such as erosion, climate change, and pollution can cause serious health problems, fertility reduction, outward migration, and capital destruction – increasing vulnerability when working in tandem with other kinds of sudden hazards. The major problem with classifying these types of disasters is that the impact of the hazard or shock can be difficult to isolate from other factors contributing to the same outcome.[8]

We can also establish a taxonomy based primarily on the consequences of disasters. Just like triggering events, consequences can be ranked according to their nature; for instance by distinguishing between demographic effects (raised mortality or reduced fertility), physical effects (destruction of land, buildings, infrastructure, and machinery), economic effects (directly as a result of physical destruction or indirectly due to

---

[5] Blaikie et al., *At Risk*, 75.   [6] Renaud et al., 'A Decision Framework.'
[7] Pereira, 'The Opportunity of a Disaster,' 469–472.
[8] Renaud et al., 'A Decision Framework', e20–e21.

erosion of livelihoods or redistribution in the long run), and social and political effects (social polarization, unrest, or upheaval). Some of the classification systems even try to form categories that incorporate two dimensions, such as the Modified Mercalli Intensity scale, which, unlike the Richter scale, incorporates human impact and building damage as well as the magnitude of a seismic shock.[9] Categories, and their relationship to disasters, are not always clear-cut, however. Demographic consequences, for instance, are obviously highly relevant in the case of epidemics, and mortality has been cited as one of the essential ways of distinguishing subsistence crises or dearth from famine.[10] However, mortality can also be prominent in geological disasters such as large earthquakes or tsunamis, or in meteorological ones. For example, the 2010 earthquake in Haiti created conditions conducive to the spread of diseases such as cholera.

The fact that disasters show such variety, both in their causes and in their consequences, raises a question vital to the core of this book: if this is the case, is it at all possible to analyze disasters using a general conceptual framework? Does it make sense to compare epidemics and earthquakes, or tsunamis and sand drifts? Overall, we believe so. While in the practice of disaster management the exact measures to mitigate the impact of a hazard or prevent its recurrence vary depending on the nature of the trigger, on a higher level of abstraction significant similarities can still be demonstrated. The ways in which individuals, groups, and societies cope with shocks – or fail to do so – often share characteristics – and this is something particularly brought to the fore when viewed in a historical perspective.

Indeed, two crucial concepts here are vulnerability and resilience.[11] Determinants of vulnerability, although situationally specific, often incorporate various aspects of distribution of wealth, resources, support, and opportunity, while resilience is determined to a significant extent by social, economic, and political institutions and the context in which they function.[12] Similarities of this type allow us to compare disasters that at first sight appear very different. For example, 'entitlement theory,' originally developed by Amartya Sen to explain vulnerabilities to twentieth-century famines in the developing world, has recently been used as a concept to assess vulnerabilities during large-scale flooding of coastal regions in the pre-industrial North Sea area and to analyze the

---

[9] Wood & Neumann, 'Modified Mercalli Intensity Scale.'
[10] Alfani & Ó Gráda (eds.), *Famine*; Ó Gráda, *Famine*, 5. See also the next section.
[11] For these concepts, see Sections 2.3.3 and 2.3.4. These concepts are employed throughout Chapters 3 and 4 in particular.
[12] Wisner, 'Disaster Vulnerability'; Van Bavel & Curtis, 'Better Understanding Disasters.'

opportunities of specific groups to organize protection against flooding and restrictions on their ability to do so. Just as weather-induced harvest failures lost their central role in explaining famines, the entitlement approach explains floods as a result of declining entitlement to flood protection by specific groups rather than by looking at storminess or climatic factors.[13]

2.2 Scale and Scope of Disasters

The scale and scope of disasters is something that continues to help them appeal to the popular imagination. By scope, we refer to the range of different facets of everyday life a disaster can touch; by scale, we refer to the intensity, magnitude, or territorial spread of effect for each of these facets touched. The diversity in scope and scale of disasters makes them a suitable subject for all kinds of popular rankings, many of which are found on easily accessible resources such as Wikipedia.

Historical research into the scale of disasters has quite a long tradition of focusing predominantly on the death toll – mortality being an important measurement for historical demographers of the 1960s and 1970s, who defined 'crisis severity' through death rates using various kinds of debated methodologies.[14] However, as disaster research has become a topic in its own right, new parameters have been added, mainly focusing on material losses, as these were highly relevant in dealing with the outcome of a disaster. Foster's calamity magnitude scale, developed in the 1970s, took into account the number of fatalities, the number of seriously injured, infrastructural stress, and the total population affected. According to this scale, the Black Death emerged as the largest 'natural' disaster, after the 'human-made' shocks of the World Wars.[15] The logarithmic Bradford disaster scale, developed for disaster prevention and management, combines fatalities, damage costs, evacuation numbers, and injuries.[16] Nowadays, physical damage is seen as essential to call an event a disaster, "because that is the perspective of institutions charged with their management."[17] Government agencies and insurers are most interested in material losses, as this is the exact aspect they have to resolve. The 1995 Chicago heat wave, for example, which killed approximately 140 people, was not formally identified as a disaster by the US government, even though the death toll was higher there than in previous years' Californian earthquakes, which were labeled as

---

[13] Soens, 'Flood Security.'
[14] An elucidation of the major early work by Goubert, 'Historical Demography.'
[15] Foster, 'Assessing Disaster Magnitude,' 245.   [16] Keller *et al.*, 'Analysis of Fatality.'
[17] Tierney, 'From the Margins to the Mainstream?,' 508.

disasters.[18] Accordingly, the definition of a term such as disaster always remains fluid and flexible and can diverge across social interest groups, even for the same event.

Although the scope of disasters can be wide, then – affecting very different aspects of social life – the main way of classifying disaster damage has still tended to revolve around casualties and material damage. And it is not always a given that the scale of effect in one dimension will be the same in the other. In some cases, a shock destroys capital but leaves people untouched, and on other occasions a shock kills people, but leaves much of the infrastructure and goods intact – the Black Death of 1347–52 being the classic example. Sometimes both occur together. Certain hazards such as floods have rarely killed large numbers of people throughout history – those that do are exceptional.[19] Other hazards such as earthquakes cause fatalities, but those deaths are then further supplemented by events occurring in the aftermath – see the already-mentioned example of cholera in post-earthquake Haiti in 2010 and 2011.

Famines and epidemics are the best example of disasters with a high death rate, and even famine mortality tends to be created mainly by conditions conducive to the spread of diseases rather than starvation per se – at least prior to the twentieth century. Classic famine-related diseases include tuberculosis, dysentery, typhoid fever, and typhus, which are linked to acute malnutrition, eating foods not normally fit for human consumption, a decline in the attention paid to hygiene, and increasing amounts of displacement and migration.[20] Plague is not said to be a disease of malnutrition, which may account for the lack of association between grain prices and plague mortality in certain early-modern studies,[21] but still may have been indirectly stimulated by breakdowns in infrastructure, poor hygiene, and heightened migration during food crises.[22] The scale of these mortality effects was not always the same, however – even with regard to the same type of disease, and to the same outbreak.

Even within the same disease, the severity and pervasiveness of outbreaks could differ significantly. The plague epidemic in 1629–30 in Northern Italy, for example, was much more severe than the plague outbreaks of the sixteenth century in the same area, and more severe than the same 1629–30 outbreak that had occurred in parts of Central Italy.[23] Many diseases killed large numbers of people in a restricted

[18] Tierney, 'From the Margins to the Mainstream?,' 508.
[19] Soens, 'Resilient Societies.'
[20] Ó Gráda & Mokyr, 'What Do People Die of during Famines.'
[21] Curtis & Dijkman, 'The Escape from Famine.'
[22] Alfani, *Calamities and the Economy*.
[23] Alfani, 'Plague in Seventeenth-Century Europe.'

number of localities, but few combined the key facets of death rate and territorial spread to be truly large-scale killers. Understanding the diverse scale of the death toll is often difficult in a historical context because (a) it requires large amounts of epidemiologic data easily comparable over long periods of time and large areas – something we do not always have, particularly before the early-modern period, and (b) the causes are immeasurably complex, bringing together interrelated factors on environment, climate, natural and acquired immunity, proximity to vectors and points of contagion, pathogen adaptation, and human patterns of warfare, migration, trade, commerce, and institutional control of the disease.

The scale of material damage stemming from disasters can also differ hugely. For some disasters it was decidedly limited: many famines, for example, had only a limited impact on capital goods, if they were not twinned with warfare.[24] Earthquakes, however, could do much more material damage, as the one that hit Lisbon in 1755 illustrates. The earthquake and the tsunami and fire that followed in its wake made two-thirds of the city uninhabitable, and destroyed 86 percent of all church buildings.[25] Yet it was not only the type of disaster, but also the society it struck that determined the scale of material damage. As an example, the indigenous Filipino way of building *nipa*'s – palm and bamboo huts – was seen as completely backward by the Spanish colonial powers, but these structures had the advantage of being easily rebuilt after earthquakes. In contrast, the seventeenth-century Spanish baroque stone buildings were reduced to ruins by an earthquake in 1645, creating much greater material damage.[26] It is telling that between 1977 and 1997 the number of deaths from 'natural' disasters remained more or less constant (even as the world population increased), but the cost of disasters increased significantly.[27] Our highly technological societies of today have become more vulnerable in terms of the specific category of material damage.[28]

The assessment of casualties and material damage is, however, a complex undertaking and historians and social scientists should note the difficulties that often arise – particularly in light of 'popular' interest in these kinds of facets of disasters. This issue can be demonstrated by looking more closely at some of the 'rankings' that often appear on the Internet – for example, the ranking of 'death tolls' from 'natural disasters' taken from Wikipedia and presented in Table 2.1. These lists, like many

[24] Gutmann, *War and Rural Life*, 3, 8.
[25] Pereira, 'The Opportunity of a Disaster,' 473–477. See also Section 6.1.2.
[26] Bankoff, 'Cultures of Disaster,' 266–267.
[27] Alexander, 'The Study of Natural Disasters,' 285.
[28] See also Sections 4.2.2 and 6.1.1.

Table 2.1 *Ranked list of natural disasters by death toll on Wikipedia, https://en.wikipedia.org/wiki/List_of_natural_disasters_by_death_toll*

Rank	Estimated death toll	Disaster	Location
1	1,000,000–4,000,000	1931 China floods	China
2	900,000–2,000,000	1887 Yellow River flood	China
3	830,000	1556 Shaanxi earthquake	China
4	≥500,000	1970 Bhola cyclone	East Pakistan
5	316,000	2010 Haiti earthquake	Haiti
6	300,000	1839 India cyclone	India
7	273,400	1920 Haiyuan earthquake	China
8	250,000–300,000	526 Antioch earthquake	Byzantine Empire
9	242,769–655,000	1976 Tangshan earthquake	China

other related ones on the Internet, tend to exaggerate the death rate for historical floods and earthquakes, sometimes even producing estimates that exceeded the total population count of the day, which then run the risk of producing sensationalist stories and narratives. What are the reasons for this? A significant problem is that many of the estimates made by contemporary observers are taken at face value, when several will have been exaggerated for a particular agenda (tax concessions, for example) or a moralizing standpoint or rhetorical effect.[29] Sometimes the guesses of contemporaries were simply that – guesses. Even in the late eighteenth century, when statistical material became more important and more prevalent in disaster-reporting in newspapers, numbers were by no means exact. Numbers are also not necessarily neutral, since they "incorporate the values of the people who create them, and data collection begins with the collector's interests or concerns."[30] Chinese victims in the 1906 San Francisco earthquake and Aboriginals in twentieth-century Australian cyclones were simply not counted – which is problematic, given that in both cases they represented numerically large proportions of the population.[31] During the Lisbon earthquake of 1755 both Protestant and Catholic commentators strongly inflated the number of dead, to fit in with their respective narrative of divine retribution for the city's godlessness. The Marquis of Pombal, the Portuguese prime minister and personification of Enlightenment and 'godlessness,' was dismayed by this and ordered his own 'official' damage report, one of the first of its

---

[29] Concern over this issue in Squatriti, 'The Floods of 589,' 820; Rheinheimer, 'Mythos Sturmflut,' 30.
[30] Aguirre, 'Better Disaster Statistics,' 29–30. See also Section 3.1.1.
[31] Aguirre, 'Better Disaster Statistics,' 29.

kind.[32] Measuring the impact of disaster is thus not as easy as Wikipedia may have us believe.

This should not be seen as merely a problem of 'popularizing' media sources either. This is because, as a result of the recent trend towards increased use of historical data by scholars working in the natural and social sciences (i.e. not trained historians), a number of papers published in high-ranking science journals are now being accepted that use figures and data taken from the Internet or historical papers, without taking into account all the methodological issues or lacunae in data collection. For a discussion on this trend, see Section 3.1.3 on source criticism and big data.

2.3 Concepts

Having explored variations in the types, scale, and scope of disasters, this section provides a critical introduction to key concepts used to study disasters – primarily those used in the disaster studies literature, but also in cognate fields such as the ecological sciences and development economics. It makes particular reference to vulnerability, resilience, and their temporal dimensions, while acknowledging that use of these concepts is inconsistent and occasionally ambiguous between disciplines and contexts.[33] Whereas in ecology, for instance, resilience is increasingly seen as the adaptive ability to transform to a different state, development economists often use a more limited definition that highlights a society's ability to return to its pre-existing state.[34]

2.3.1 Disaster and Hazard

One point of contention in the disasters literature lies in the term 'disaster' itself, and how it should be defined – and distinguished from other terms such as 'catastrophe' and 'shock.'[35] Certainly there is a tendency towards separating qualifiers such as 'natural' from the term 'disaster.' Although the term 'natural disaster' has fairly widespread use as a convenience term in the mainstream media and some popularizing literature, few scholars now argue that disasters are simply 'natural events,' regardless of whether they are working in the natural sciences, social sciences, or humanities. Indeed, it is clear that although hurricanes, for example, are fundamentally natural phenomena, the

---

[32] Aguirre, 'Better Disaster Statistics,' 33.
[33] Kelman *et al.*, 'Learning from the History.' [34] Sudmeier-Rieux, 'Resilience.'
[35] The origins and etymology in a historical perspective are described at length in Mauelshagen, 'Defining Catastrophes'.

root cause of disastrous effects emerging from hurricanes can usually be put down to poor building construction or weak institutional infrastructure rather than the occurrence of extreme winds or storm surges per se. Indeed, the past few decades of disaster studies research have consistently shown that it are social processes that shape disasters and those most at risk. These include technological, political, and cultural factors that determine human capacity to prepare for, cope with, and recover from sources of potential harm,[36] as well as gender, ethnicity, and age – each of which has little to do with the natural environment.[37] The emphasis on the *naturalness* of disaster that comes with the term 'natural disaster' focuses attention on physical processes and their destructive power, rather than what makes people vulnerable to these processes, and can distract attention from human responsibility for the causes of disasters.[38]

Attention in the disasters literature has instead turned to terms such as environmental 'hazards,' or sometimes 'shocks.' The word 'shock,' with its inbuilt element of surprise, does not hold the same breadth of applicability as hazard, as it implies that the event or process was unexpected. This may be the case in an area suffering a severe tsunami triggered by a distant high-magnitude earthquake, but the same may not be said for a river bursting its banks onto a floodplain. Still, the words 'shock' and 'hazard' more aptly describe an environmental event or process itself, whereas 'disaster,' or perhaps 'nature-induced disaster,' refers specifically to the severe impact of an event.[39] As we have seen, hazards can be both natural and technological, and both can lead to disasters. But before we examine the factors that might turn a hazard into a disaster, how do we distinguish a disaster from a hazard in the first place? This question has been the subject of much debate in disaster studies.[40] Previously we showed the complexities of classifying disasters on the basis of characteristics such as magnitude, duration, impact, potential of occurrence, and ability to control impact,[41] as well as consequences such as economic losses and mortality.[42] In reality, any such distinction or threshold is an inherently anthropocentric valuation, and any metric is open to criticisms of generalization. What is clear, though, is that there is a qualitative difference between disasters and hazardous events:

---

[36] Such factors could include where people live and work, and their wealth, health, and access to information: Pelling, *The Vulnerability of Cities*.
[37] Wisner et al., *At Risk*.   [38] Ribot, 'Cause and Response.'
[39] Cohen & Werker, 'The Political Economy.'
[40] Perry & Quarantelli, *What Is a Disaster?*; Quarantelli (ed.), *What Is a Disaster?*
[41] Berren, Beigel & Barker, 'A Typology.'
[42] Foster, 'Assessing Disaster Magnitude'; Keller, Wilson & Al-Madhari, 'Proposed Disaster Scale.'

disasters severely disrupt normal activity, often cause damage or casualties because coping capacities are exceeded, and require large responses in terms of resources and organization which often necessitate external support.

If an extreme geophysical or meteorological event is not simply a synonym of a disaster, and disaster risk is produced by a combination of factors, it becomes imperative to understand the determinants of different levels of *vulnerability* of different groups of people and how societies tried to *manage* hazardous events.

2.3.2 The Disaster Management Cycle

Some disasters have also been classified according to how they are managed, and this approach is seen nowhere more clearly than in the disaster management cycle – a framework to understand the processes and stages through which disasters evolve. This framework has been used, further developed, and modified in a variety of disciplines including sociology, geography, psychology, civil defense, and development studies. The initial idea to develop a framework to understand disasters via how societies cope with their effects dates back to the 1930s. Practitioners and policy makers distinguished between different phases of the unfolding disaster to respond more effectively in future situations. Initially, three stages were identified: a preliminary stage where the hazard and problems built up, followed by the disaster stage in which the actual event took place, and finally the readjustment or reorganization stage.[43]

In the 1970s the idea of a cycle was developed because of the often-recurrent nature of hazards and the disasters that can ensue. Seldom is a society hit by a completely unforeseen hazard or disaster, such as a meteorite or sudden earthquake in low-tectonic-risk zones. A series of disasters occurring during that decade urged practitioners and policy makers to look for more than simply disaster relief measures: societies had to become more receptive to prevention. Therefore, the disaster management cycle was developed, comprised of four phases including mitigation after a previous disaster, the development of preparedness, the response after another triggering event, and recovery.[44]

Since then different disciplines and scholars have proposed several models with more or fewer stages. Here we present the disaster cycle as recently reformulated by John Singleton as an analytical tool for

---

[43] Coetzee & van Niekerk, 'Tracking the Evolution.'
[44] Singleton, 'Using the Disaster Cycle.'

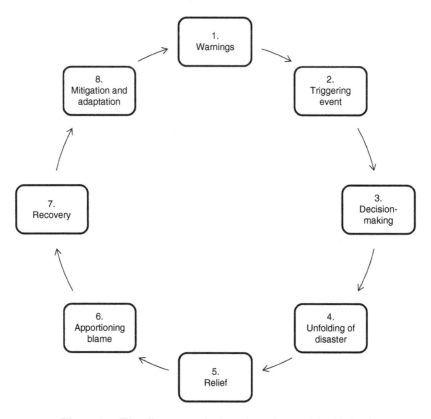

Figure 2.1 The disaster cycle, based on the model of John Singleton. Singleton, 'Using the Disaster Cycle.'

historical research (Figure 2.1), which includes more psychological and social processes than the 1970s framework. We use this disaster cycle as an illustrative example precisely because it fits within our framework for two reasons. First of all, the model acknowledges that hazards do not happen out of the blue but interact with societies that have a long prehistory with similar hazards, and therefore former mitigation and adaptation measures affect the outcome of a disaster. Second, the cycle moves beyond the short-term disaster effects and mitigation stages, and urges scholars to look at the long-term effects and adaptation measures. Therefore, this model is conveniently set up to address the temporal components of historical disasters. In this framework, decision-making is one of the most crucial stages between the occurrence of hazard and potential disastrous consequences, and thereby a driver of high or low

2.3 Concepts

disaster impact. Those intermediate response and decisions stages have also been described as 'sensemaking' points.[45]

Although providing a more 'complete' overview of how disasters can unfold from early warning signs and triggers, to adaptive responses, and then the aftermath of recovery, there are also some limitations to classifying and conceptualizing disasters in a full cycle. It must be remembered that this is a disaster *management* cycle, not a disaster cycle; and, furthermore, it is definitively a cycle and thus connected to cyclical processes only. Many disaster 'cycles' do not reach those later stages at all, and it is unclear whether all stages are actually intrinsic parts of the disaster experience. For example, according to Singleton's framework, almost all disasters lead to the apportioning of blame, and yet recent historical research has shown that even some of the most severe epidemics did not necessarily lead to scapegoating or social unrest,[46] but instead gave rise to cohesive or even compassionate responses, and some of the most severe famines did not necessarily bring about the collapse of collective solidarity mechanisms.[47] Does the absence of blame then preclude a whole host of crisis conditions from being labeled as disasters? A cycle such as this one also has no room for divergence. We are told that intermediate decision-making processes remain vital to disaster impact, but there is no possibility of measuring the effectiveness of these decisions and relief phases. Accordingly, the disaster management cycle approach to disaster classification and conceptualization offers a step forward in outlining a 'textbook case' of how a disaster unfolds over a period of time, but we need more flexible frameworks if we want to conceptualize disaster experiences that diverge from the 'norm.' For that we must turn to other concepts such as vulnerability, resilience, and adaptive capacity.

2.3.3 *Vulnerability*

It is easy to associate disasters with 'nature' when we consider that earthquakes are caused by shifting tectonic plates, floods by storm surges, and plague by biological pathogens. Nevertheless, the naturalness of natural disasters was already questioned in the 1970s in a seminal article in *Nature*, where it was stated that "the time is ripe for some form of precautionary planning which considers vulnerability of the population as the real cause of disaster – a vulnerability which is induced by socio-economic conditions."[48] Without denying that an initial hazard is often required for a disaster, this approach emphasizes that the root cause of

---

[45] Weick, 'The Collapse of Sensemaking.'   [46] Cohn, *Epidemics*.
[47] Slavin, 'Market Failure.' See also Section 6.1.4.
[48] O'Keefe, Westgate & Wisner, 'Taking the Naturalness,' 567.

a disaster is social, and that people have to be already vulnerable to hazards in order for a disaster to arise: that is to say, all vulnerability is social vulnerability.[49] Through this lens, the most important questions are therefore twofold: who suffers from disasters, and why does society create these precarious circumstances that expose people to suffering?

Before addressing those questions, however, a more essential question is what do we mean by vulnerability? According to one influential work from the 1980s, vulnerability was meant to represent "the exposure to contingencies and stress and difficulties coping with them," and is comprised of two sides: an external side of risks, shocks, and stress to which an individual, household, or community is subject, and an internal side concerning a lack of means to cope.[50] According to this kind of definition, the concept then rests on three distinct facets: (1) a risk of exposure; (2) a risk of inadequate capacities to cope; and (3) the attendant risks connected to poverty.[51] These facets of vulnerability can be found within individuals, but also on the collective level of social groups, communities, and whole societies.

The above-mentioned definitions and general approach to vulnerability were at their most dominant in the 1970s. On the basis of critiques of Western development plans for the Global South, this movement stated that vulnerabilities arose from political choices rather than from natural inevitabilities. Its popularity, however, waned in the decades thereafter. Driven by increasing insight into climate change and its human components, systemic approaches began to gain ground (focusing on ecosystems, for example), in which the ability of systems to absorb, adapt, and transform when confronted with disasters was considered pivotal. In particular, much of the focus connected to climate change shifted from different and diverse social groups to either the 'system' as a whole or the individual, and hence from vulnerability to resilience (see the next section). Social relations and the application of power were no longer central to many of the hazards and disasters narratives, as attention moved from causality to response.[52] In the process, as the 'social' aspect of disasters became increasingly obscured, so too did the structural inequalities in wealth, resources, and power that shape how disasters impact societies and people differentially. This is why several authors linked this shift and the new focus on resilience and adaptation to the hegemony of neoliberal ideologies.[53]

---

[49] Most famously argued in Blaikie et al., *At Risk*, 7, 9.
[50] Original definition in Chambers, 'Vulnerability, Coping and Policy,' 1.
[51] Watts & Bohle, 'The Space of Vulnerability,' 45.
[52] Ribot, 'Cause and Response,' 669.
[53] Cannon & Müller-Mahn, 'Vulnerability, Resilience and Development Discourses'; MacKinnon & Driscoll Derickson, 'From Resilience to Resourcefulness'; Sudmeier-Rieux, 'Resilience.'

Nevertheless, recent work has argued for disaster research to once again return to vulnerability as a core organizing concept. This argument centers on the potential of vulnerability to put the 'social' back into disaster analysis, focusing on elements that run the risk of being neglected in more instrumentalist or technocratic approaches to disasters. Indeed, by placing particular attention on the root causes that make people vulnerable, we are able to shine more light on different aspects that become apparent only over long periods of time, and by moving beyond directly disaster-related issues.[54] For those looking to use history as a tool for understanding more about different dimensions of disasters, this inherent temporal aspect of the vulnerability approach becomes invaluable – whether working across years, decades, or centuries – since the vulnerability of individuals, groups, and communities varies over time, and we need the frameworks to analyze and understand these changes.

2.3.4 Resilience

While disaster studies scholars from the 1970s to the 1990s were preoccupied with vulnerability and its root causes, resilience became the buzzword of disaster studies at the start of the twenty-first century. The origins of resilience – from the Latin *resilire* meaning more literally to 'jump back' – can be traced back to the 1940s and 1950s when the concept was used both in psychology ('lives lived well despite adversity') and in engineering (the capacity of materials to absorb shocks and still persist). Yet it was from ecosystem analysis that the concept migrated to disaster studies. As defined by Buzz Holling in 1973, resilience refers to either the 'buffer capacity' of an ecosystem (its ability to absorb perturbation), the magnitude of the disturbance that can be absorbed before structural change occurs, or alternatively the time it takes to recover from disturbance.[55] Later, the concept was transferred to the social sciences. W. N. Adger defines it as "the ability of communities to withstand external shocks to their social infrastructure" and sees a direct link between 'ecological' and 'social' resilience – particularly in societies highly dependent on a single resource or a single ecosystem.[56] Over time, fostering resilience became the official mantra of international disaster relief and prevention, with the overarching idea that if we can strengthen the capacity of households and communities in risk-prone areas to counter hazards, for instance by improving alert systems and

---

[54] Bankoff, 'Remaking the World.'
[55] The pioneering article was Holling, 'Resilience and Stability'; see also Adger *et al.*, 'Are There Social Limits,' 349.
[56] Adger *et al.*, 'Are There Social Limits,' 361.

solidarity networks, organizing micro-credit, or removing institutional constraints to food markets, then we can accommodate recurrent hazards.

While the initial focus of social science research into resilience was about 'bouncing back' after disasters, things became complicated with the realization that change following a shock is not necessarily something that can be viewed negatively, since it can also stimulate changes for the better. In disaster studies, older 'conservative' definitions of resilience – measuring the restoration of the previous equilibrium – became replaced by more 'progressive' ones, seeing adaptation of the system in more positive terms.[57] When we look to the past, there are some examples of that too. In the fourteenth century, climatic cooling and epidemic diseases may have resulted in the retreat of settlement and arable cultivation in upland England or Scandinavia – on the surface a 'negative' outcome – and yet this could also be interpreted as simply rearranging farming and habitation into more fertile areas, or a complete shift in production from arable to pastoral.[58] Apart from material or demographic changes, institutions could be transformed as well to suit the new environmental and societal structures that develop after a shock.

By stretching this idea too far, however, new problems can be created: if the complete make-over of a society after a major disaster is qualified as a 'resilient' outcome, then only total breakdown or collapse remains as counter-evidence for a failure in resilience. And although popular books have been written on the subject of collapse of societies and civilizations in the past,[59] from a historical perspective we are also aware that total breakdown and collapse has been exceptionally rare – certainly over the period for which written documents survive. Accordingly, it might also be the case that most studies on past hazards and disasters reach the same conclusion as Georgina Endfield in her thought-provoking discussion of extreme drought and floods in colonial Mexico: society did not collapse, but proved 'remarkably resilient' to such problems.[60] If hardly anything can counter the resilience outcome – and history proves that to be the case more often than not – then the term begins to lose much of its utility.

As noted towards the end of the previous section, further critiques of resilience include the subordinate role given to power relationships, agency, values, and knowledge. Some authors even see resilience as the handmaid of neoliberalism, strengthening its discourse on personal responsibility, but this is a "responsibility without power."[61] This is

---

[57] Endfield, 'The Resilience and Adaptive Capacity,' 3677.
[58] Dyer, *Standards of Living*, 259–260.   [59] Diamond, *Collapse*.
[60] Endfield, 'The Resilience and Adaptive Capacity,' 3677. See also Section 5.2.
[61] MacKinnon & Driscoll Derickson, 'From Resilience to Resourcefulness,' 255.

2.3 Concepts

particularly problematic given that systems that may be classified as highly resilient can contain, or even derive some of their resilience from, vulnerability within certain groups or communities.[62] Ultimately in this book, we accept the more recent definition of resilience as something systemic, where adaptation can lead to a new post-hazard or post-disaster 'state of things' (see the next section on adaptation), but we also more explicitly demonstrate our criticisms of the concept by employing historical examples in Chapters 5 and 6. In the end, while a certain level of resilience becomes the outcome from most historical disasters, vulnerability outcomes are much more diverse and unpredictable.

2.3.5 Adaptation, Transformation, and Transition

Adaptation generally refers to "the adjustments that populations take in response to current or predicted change," and is related to each of the frameworks introduced above.[63] Emphasis on adaptation was long shunned within the Intergovernmental Panel on Climate Change (IPCC) in favor of the mitigation of greenhouse gas emissions, whereas acceptance of a widespread need for adaptation was seen as accommodating (or even embracing) the inevitability of disaster.[64] Over time, however, the term adaptation or adaptive capacity did gain currency – particularly from the early 2000s – and in close parallel with the dominance of more resilience-focused research that accepted more flexible definitions of resilience outcomes. This was initially very prominent in climate change research, where models within the natural sciences prescribed technological 'fixes' for issues such as declines in agricultural productivity.[65] Only later did social science approaches refocus research into adaptation onto areas such as indigenous knowledge – but also cultural limits to adaptation and even the potential negative consequences of adaptive action, which became known as 'maladaptation.'[66] Thus, we came to learn that hazards and disasters often led to adaptation, with adaptive capacity and resilience closely linked,[67] and this has had the knock-on effect of allowing us to exchange gloomy interpretations of disasters for more 'positive' ones, stressing the opportunities for change

---

[62] Cannon & Müller-Mahn, 'Vulnerability, Resilience and Development Discourses.'
[63] Nelson, Adger & Brown, 'Adaptation to Environmental Change.'
[64] Pielke Jr. et al., 'Climate change 2007,' 445, 597–598; Kaika, *'Don't Call Me Resilient Again!'*; O'Connor et al., 'Living with Insecurity'; Reid, 'The Disastrous and Politically Debased Subject.'
[65] Noble et al., 'Adaptation Needs and Options.' See also Pelling, *Adaptation to Climate Change.*
[66] Barnett & O'Neill, 'Maladaptation'; Adger et al., 'Are There Social Limits'.
[67] Engle, 'Adaptive Capacity.'

created by the disaster.[68] In more recent times, historians have argued for cases where climate-related pressures, leading to hazards, did indeed create new incentives for adaptation, since risks and hazards could function as constructive triggers for innovations. See for example the economic and cultural flourishing of the Dutch Golden Age of the seventeenth century being tied to successful adaptive capacity during the worst phases of the so-called Little Ice Age.[69]

However, at the same time, we have also come to learn that adaptation did not always occur post-hazard or post-disaster, and equally that not all of the adaptation that did occur can be considered in 'positive' terms. Elements such as social capital, networks, trust, and coordination are often cited as factors promoting adaptation,[70] yet elements hampering adaptation have been cited, such as mismatches in scale between environmental and social dynamics, asymmetries in power, and inequality.[71] Much of this is further complicated by the intricacies of scale: hazards that lead to disastrous disruption at the level of single-village communities may be perfectly absorbed on a regional or macro-level, for example. Adaptation, and the form it takes then, is never inevitable – and a path forward for understanding how societies respond to hazards and the disasters that ensue is surely connected to better understanding why certain systems and societies adapt and why some do not, and, moreover, why some adaptations are effective, while others are less so. As Eleonora Rohland has noted, this perspective is seemingly at odds with definitions of adaptation that include notions of "moderating harm" or "exploiting beneficial opportunities" – both of which depend on time, place, conflicting interests, and power relations[72] We suggest that this is an area in which historians can offer the greatest insights and contribution, since an important element for explaining adaptation is incorporating chronology and developments across time.

Indeed, influential models for adaptation of social systems or ecosystems – such as the 'Adaptive Cycle' of Gunderson and Holling – pay explicit attention to temporal aspects and the different phases in which adaptation can take place, but stress that this adaptivity has its limits when confronted with hazards that are too numerous or too extensive.[73]

---

[68] For disasters as opportunity, see Section 5.3.1. [69] Degroot, *The Frigid Golden Age*.
[70] Adger, 'Social Capital'; Barnes et al., 'The Social Structural Foundations'; Bodin, 'Collaborative Environmental Governance.'
[71] Cumming, Cumming & Redman, 'Scale Mismatches'; Crona & Bodin, 'Power Asymmetries'; Van Bavel, Curtis & Soens, 'Economic Inequality'; Cumming & Collier, 'Change and Identity in Complex Systems.'
[72] Rohland, 'Adapting to Hurricanes.' These terms are included in the IPCC definition of adaptation.
[73] Gunderson & Holling, *Panarchy*, esp. 34; see also Scheffer, *Critical Transitions*; Folke et al., 'Resilience Thinking.'

2.3 Concepts

According to this framework, adaptation within socio-ecological systems accommodates external disruption as easily as internal dysfunction, but over time the system becomes more rigid and difficult to adapt – leading to a lack of flexibility and 'rigidity traps' – which in turn creates a kind of tipping point or threshold that can lead to implosion from even the smallest of shocks or slightest disruption. Historians have indeed shown – for example in the case of success or failure in adapting to hurricanes – that rigid configurations of technologies or institutions can be difficult to transform, even in the wake of severe hazards.[74] More broadly speaking, recently support has gathered for historically informed approaches towards climate change adaptation, with greater attention paid to long-standing or even path-dependent norms and processes that drive or constrain successful or unsuccessful adaptation in particular social contexts.[75]

Therefore, through a lens with more attention paid to temporality, historians can fundamentally complement and alter the current focus on adaptation. For one, historical research may help to redefine practices that on the surface appear 'maladaptive.' For example, it has been shown that the migration of pastoralists in times of drought should not be conceived negatively but instead offers long-standing and effective ways of sustaining livelihoods in the face of climate-related hazards.[76] Conversely, historical research may dampen current optimism about the possibility for adaptation. While ecosystems can perhaps indefinitely and automatically adapt, historians by investigating adaptation processes in the long term can show how human societies do not have the same logic. Some did adapt, while others maintained a rigid, ineffective, or even destructive institutional framework. By examining how adaptation and the problems to which we are adapting emerge over time, it also becomes possible to identify the level of action needed to reduce pre-existing vulnerabilities – whether this concerns incremental changes over extended periods or transformational change in the face of deep-rooted and recurring problems.[77]

2.3.6 Risk

So far we have focused on physical exposure to natural hazards, and more significantly issues with societal organization that lead to pre-existing vulnerabilities of certain populations, yet another fundamental element

---

[74] Rohland, 'Adapting to Hurricanes.'
[75] Adamson, Rohland & Hannaford, 'Re-thinking the Present.'
[76] Agrawal & Perrin, 'Climate Adaptation.' [77] Pelling, *Adaptation to Climate Change.*

for understanding the occurrence and impact of disasters is 'risk.' Risk relates to human agency and perception, which guide the strategies deployed by individuals or groups to manage and calculate the potential occurrence of harm. The calculation of risk is often based on weighing up the possibilities, likelihood, and consequences of a number of outcomes, and preference for an outcome differs from context to context depending on the values of the parties involved – that is to say, it is highly subjective. So, in many pre-industrial contexts, for example, societies had to weigh up the risks of living in a particular environment. Some agricultural workers lived in agglomerated settlements – even when they were poor, cramped, and far from their fields – because to live isolated next to their lands exposed them to the risks of violence.[78] But this also depended on the weight of knowledge. Some people in late-medieval and early-modern Europe moved to cities – to find work and have access to urban amenities – but at the same time heightened risks of death through disease outbreaks – the so-called 'urban penalty,' yet, of course, ordinary people did not have equal knowledge about the likelihood of either outcome.

Time is also an important aspect of risk. According to Ulrich Beck, for example, risk has taken on whole new dimensions and meaning from the second half of the twentieth century – and is thus not directly comparable to earlier risks. Indeed, the increasing number of 'technological' disasters (from the Bhopal gas tragedy to the nuclear disaster in Chernobyl), and the obvious failure to predict and control natural variability and climate change, has apparently introduced a new kind of 'risk society,' where perfect control has been abandoned, but the management and accommodation of uncertainty remain center-stage.[79] This shift in risk perception is reflected in the evolution of private insurance schemes related to climatic risks: expanding during the twentieth century, but declining again at the dawn of the twenty-first, as the potential losses were increasingly deemed impossible to cover by private insurance schemes.[80]

Risk used in this way can also be critiqued in two ways. First, some works have argued that risk is not a neutral concept, but that its use is determined by inequitable power relations. According to disaster scholars such as Greg Bankoff, the so-called Third World has been significantly 'othered' throughout time by its repeated associations with risk. During colonial times, colonized countries were seen as disease-ridden and in need of a civilization offensive. After the World War II, focus was put on remedying poverty, but during the 1990s attention shifted to the

---

[78] For example in Southern Italy: Curtis, 'Is There an Agro-town Model.'
[79] Beck, Lash & Wynne, *Risk Society*. [80] Andersson & Keskitalo, 'Insurance Models.'

'disaster-proneness' of non-Western countries, which were labeled as risky environments and areas of disaster. In a way, our labeling of these regions as hazardous and risky also demonstrates a clash between two different types of risk societies: the one where risk as a frequent life experience has led to adaptation (as in several non-Western cases) and the one where risk was seen as needing to be controlled and later managed (as in the Western world).[81]

Second, it is our view also that the perception of modern risk being inherently or intrinsically different from that of the pre-modern era is perhaps overstated slightly. Although perfect control may have been abandoned as a disaster philosophy in recent times – with the accommodation of uncertainty now more acceptable – it is clear that very few pre-industrial societies believed they could completely eradicate the possibility of experiencing both hazards and disasters. For many of them, the constant threat of certain hazards, and the management of their prevention and impact, was a central preoccupation, with the acceptance of natural hazards as 'frequent life experience' and continuous attempts to adapt both landscape and society to accommodate risk as well as possible.[82] Epidemics – once said to evoke total panic and breakdown – by the early-modern period at least became simply accepted as 'normal' characteristics of urban life, and while precautionary measures were developed, social transactions did not stop altogether. Peasant societies are often a prime example of this – both pre-industrial and contemporary – by shaping their survival strategies to deal with risks inherent to their way of life, but never eliminating them. In a much-cited article from 1976, D. McCloskey isolated this attitude towards risk as the distinctive element of subsistence-oriented societies.[83] Risk aversion characterized subsistence-farmers, who in order to guarantee the long-term survival of the family, tended to diversify income, crops, and plots, preferring a stable but low income to higher but less certain profit.[84]

Still, however, it is important to point out that the 'risk society' model of hazard and disaster behavior is not seen in all occasions of the past. Commercial and capitalist societies in Europe, on the rise during the early-modern period, saw the development of a new and rational risk paradigm: one founded on a belief in the capacity of humans to control the 'vagaries' of nature, and based on the development of technological, financial, and institutional 'improvements.' Risk became a rational operation which

---

[81] Bankoff, 'Rendering the World Unsafe,' 31.
[82] Bankoff, 'The "English Lowlands,"' 19.
[83] McCloskey, 'English Open Fields'; also McCloskey, 'The Prudent Peasant.'
[84] Pretty, 'Sustainable Agriculture.'

could be calculated, predicted, and hence controlled.[85] In the modern tradition, risk became equated with opportunity, a sign of human progress. Originating in international trade, insurance contracts became the main institutional tool of dealing with nature-induced risks.[86] Accordingly then, we suggest that risk retains its place as an important concept for disasters in history – since, on the one hand, we should refrain from going as far as Beck to conceive of the distant past and the contemporary as two different and incomparable 'worlds,' and yet, on the other hand, we should recognize that risk could mean very different things over time – with its meaning often dictated by changes in the social distribution of resources and power.

[85] Lübken & Mauch, 'Uncertain Environments,' 2–4; Jaeger *et al.*, *Risk, Uncertainty, and Rational Action*; Giddens, *The Consequences of Modernity*.
[86] Mauelshagen, 'Sharing the Risk of Hail'; Rohland, 'Earthquake versus Fire.'

3   History as a Laboratory: Materials and Methods

A central premise of this book is that we can use history as a 'laboratory' to test theories with relevance beyond particular time–space contexts. This is an *analytical* approach to history, one where the goal is not simply to tell the story of the past, to describe conspicuous events, or to construct time series of certain phenomena, but rather to develop and test hypotheses.[1] The study of disasters lends itself particularly well to this end for three main reasons. First, hazards, disasters, and their effects are generally well documented in historical written records across the globe, allowing us to trace their social, economic, and cultural dimensions over time. Second, environmental hazards occur at multiple scales – both spatial and temporal – and are met with divergent responses and impacts across these scales, allowing us to make comparisons and hence offer a counterpoint to the limitations of descriptive analysis.[2] Third, where we lack written information on the hazards themselves, independent of their impact, we can use other forms of knowledge such as scientific proxies as a baseline.

Using history as a laboratory to better understand disasters, however, makes practical demands of us as historians: we need suitable measures and methods to understand hazards and their effects, we can work only with what is physically available to us, and we must do so without losing sight of the critical approach to sources that forms the cornerstone of sound historical scholarship. This chapter discusses these issues: the process of identifying and interpreting sources, of reconstructing and measuring disasters, and of analyzing them in the historical laboratory. The rest of the chapter is split into two parts: the first deals with sources of data and the second with methods.

---

[1] Van Bavel & Curtis, 'Better Understanding Disasters.'   [2] See Section 3.2.3.

3.1 Historical Sources

3.1.1 Types of Historical Sources

The types of historical sources available to research hazards, disasters, and their aftermath are deeply intertwined with the characteristics of the societies producing them. This book is mainly interested in historical *written* records. This is not to say that non-written sources are irrelevant for the study of disasters and history – far from it – but rather that written sources provide the qualitative and quantitative basis for investigation into the kind of questions that our historical laboratory aims to address.[3] This section will introduce the types of sources available prior to the widespread instrumental recording of environmental variability as well as modern instrumental records. While we are concerned first and foremost with the hazard–society nexus, we also cover the types of sources and datasets available to provide information on hazards. Indeed, an almost ubiquitous feature of the historical record, at least in the pre-industrial period, is that the very same sources hold information on both hazards and their societal impacts – as outlined in Table 3.1 – making it difficult to separate one from the other in such a discussion.

For the period before instrumental recording, evidence on hazards and disasters is based on a combination of direct observations and descriptions from contemporaries, and indirect recording of processes and phenomena that were influenced by environmental conditions. This evidence was recorded in a wide range of documents of a narrative and administrative nature, although common to both direct and indirect evidence is the fact that many records were not kept for the primary purpose of systematically recording hazards and disasters but were the product of other (often economic) purposes.[4]

'Direct' documentary source types, where available, should represent the first port of call, although in practice these sources are often not widely available or detailed enough to assemble a comprehensive picture of hazards and disasters. One direct source that has received much attention is English manorial accounts extending back to the early thirteenth century CE (1230 in the case of the Bishopric of Winchester). These sources not only list the quantity of seed and yield for wheat but also make consistent reference to the influence of severe weather conditions on farming activity, providing a continuous series of seasonal climatic

---

[3] For the relevance of non-written sources, see for example Section 3.1.2.
[4] For a good indication about source types, see Pfister & Brázdil, 'Social Vulnerability'; Nash & Adamson, 'Recent Advances.'

Table 3.1 *Historical documentary evidence for reconstructing hazards and their impacts prior to instrumental recording*

Hazard	Associated impacts	Key sources
Precipitation, floods, drought	Harvest failures/shortfalls, damage to structures, loss of capital goods, malnutrition, mortality	Harvest accounts, phenological accounts (e.g. dates of flowering), rogations, burial records, colonial governmental records, missionary accounts, newspapers, private diaries (e.g. of weather, farming, or more indirect), ships' logbooks, grain prices
Temperature, ice/snow cover	Harvest failures/shortfalls, malnutrition, mortality	Harvest accounts, missionary accounts, ships' logbooks, ice-break accounts, burial records
Hurricanes, cyclones, typhoons	Inundation of land, harvest failures/shortfalls, damage to structures, loss of capital goods, malnutrition, mortality	Missionary accounts, colonial governmental records, private diaries, ships' logbooks, chronicles, gazetteers
Earthquakes, volcanic eruptions, tsunamis	Damage to structures, loss of capital goods, mortality	Chronicles, gazetteers, petitions, historical catalogs (chronologies of events)
Sand drifts, erosion, landslides	Land degradation, damage to structures, loss of capital goods	Rent and tax registers, charters, bylaws, land books, maps, reports, eye-witness accounts, tenant contracts, chronicles, petitions, newspapers
Epidemics	Mortality, reduction in fertility	Burial records, mortmain accounts, wills and testaments, ordinances, city accounts, bills of mortality, plague house/hospital documentation, medical treatises, religious tracts, orphanage records, chronicles

conditions and their agrarian impacts until the middle of the fifteenth century.[5] While these sources are invaluable, and indeed have spurred much work on the climate–society nexus in medieval England,[6] it is rare to have such a direct, consistent, and detailed source at our disposal at that point in time.

---

[5] Titow, 'Evidence of Weather'; Titow, 'Le climat.'
[6] Campbell, *The Great Transition*; Slavin, *Experiencing Famine*.

Scholars are therefore more frequently drawn to sources such as chronicles, which are much more widely available and at a larger geographical scale – particularly for medieval Europe, but also for early-colonial settings during the sixteenth and seventeenth centuries. Chronicles are books that contain a chronological narrative centered on notable occasions and often mention extraordinary weather events, diseases, and food crises. In some cases these sources represent the only narrative evidence we have on certain events. Secondary works that compile references to hazards and disasters have relied extensively on chronicles, a well-known example being Jean-Noël Biraben's work that compiles 'mentions' of medieval plague outbreaks.[7] Comparable in nature, but focusing on a specific type of event, are historical catalogs (chronologies of unusual events and their impacts) composed by contemporaries. In Italy, for instance, the first earthquake catalog, based on eye-witness accounts and information from earlier chronicles, was compiled in the late seventeenth century.[8] Other compilations assembled more recently but based on similar source types include the Mediterranean tsunami catalog, which covers the coast of Greece, Turkey, Syria, Israel, and the Southern Balkans, and extends back into the second millennium BCE.[9]

Parallels in non-Western societies sometimes go back to a more distant past and are frequently more detailed than European chronicles and catalogs. A famous example are the Egyptian Nilometers, specifically constructed to measure the heights of the floods, which were essential for agriculture. These give us an insight into the occurrence of floods and droughts. Roman examples were preserved at Aswan and Luxor; a medieval one can be seen in Cairo.[10] The Babylonian Astronomical Diaries, another example, not only record unusual natural phenomena, but also give detailed price quotations, allowing analysis of the impact of, for instance, locust invasions on markets in the fourth century BCE.[11] For China, local gazetteers – recordings of regional history and geography – form a valuable source of information. The first gazetteer dates from about 2000 years ago, but, especially during the Ming and Qing periods, thousands of gazetteers on the provincial, prefectural, and county level were compiled, usually by government officials or local scholars. Most gazetteers contain sections devoted to extreme events such as storms and floods; the later ones also give detailed information on the social, political,

---

[7] Biraben, *Les hommes et la peste*, II. See Section 3.1.3 for an analysis of the pitfalls of such datasets.
[8] Rohr, 'Man and Natural Disaster,' 130.
[9] Maramai, Brizuela & Graziani, 'The Euro-Mediterranean Tsunami Catalogue.'
[10] Shaw, *The Oxford History of Ancient Egypt*, 420.
[11] Pirngruber, 'Plagues and Prices: Locusts.'

3.1 Historical Sources

and economic consequences of such events. Data from gazetteers have, for instance, been used to reconstruct the effects of gender and family relationships on coping strategies during the North China Famine of 1876–79.[12]

Such sources, however, are mostly limited to Western and Central Europe, the Middle East, and East Asia, and elsewhere we are forced to look to other source types. For the pre-industrial period, written source availability is often greatest in areas with histories of colonialism that extend back beyond the late nineteenth century. Thus, recent work on disasters and history of a more global scope has included Southern Africa,[13] South Asia,[14] the Caribbean,[15] and South America.[16] One source that has proven to be of high value in these areas is the records of missionary societies. Missionaries were usually stationed in one area for a significant length of time and were assiduous recorders of the physical environment as it was crucial to subsistence and transportation, making these documents of particular value for the study of hazards. Moreover, unlike traders and many early-colonial officials, missionaries usually turned their attention beyond the workings of the colonial machine and onto the local population – especially during times of stress.[17] These sources have been supplemented by newspapers, which sometimes provided regular reference to provincial weather conditions,[18] and the diaries of hunters, travelers, traders, and explorers, who often reported weather conditions and their perceived impacts on the societies through which they traveled.[19] All of these sources nevertheless have their own particular biases and must be 'read against the grain' if we are to identify the local voices within the narrative. Changes in source coverage must also be taken into consideration when analyzing developments over the long term. Gaps and silences in colonial records occur for many reasons, and it must be ensured that absence of evidence relating to disasters is not conflated with evidence of absence.

Indirect evidence (or proxy data), often derived from serial administrative sources, also allows us to explore the occurrence, course, and consequences of pre-industrial disasters. In parts of Central Europe, for example, the beginning of the grape and rye harvest was reported each year to the owner of the tithe to facilitate the monitoring and

---

[12] Edgerton-Tarpley, 'Family and Gender in Famine.'
[13] Hannaford, 'Long-Term Drivers'; Nash et al., 'Seasonal Rainfall Variability.'
[14] Adamson & Nash, 'Documentary Reconstruction'; Adamson & Nash, 'Long-Term Variability.'
[15] Berland & Endfield, 'Drought and Disaster.'
[16] Prieto & Herrera, 'Documentary Sources from South America.'
[17] See for example Hannaford, 'Pre-colonial South-East Africa.'
[18] Nash et al., 'Seasonal Rainfall Variability.'   [19] Adamson, 'Private Diaries.'

collection of the crop. The close association between harvest dates and seasonal climatic conditions, however, enables these sources to be used as climatic proxy data.[20] Evidence on yields can be used as an indicator of food availability and is relatively widely available for Western Europe from the fifteenth century onwards. Price series, usually of dominant bread grains such as wheat or rye, are also frequently used as an indicator for food crisis and famine. One needs to tread carefully when using price series to reconstruct periods of dearth, however, as they are usually limited to more commercial (urban) regions and drawn from institutional accounts, whose prices are not necessarily representative of the rates at which the majority of the population acquired grain.[21] Accounts (of cities, villages, but also of religious institutions) can also contain references to extreme weather events and/or disease. In Spain and Italy, for example, the Catholic Church organized rogation services (*rogativas*) in an attempt to bring an end to situations of protracted wet or dry conditions which adversely affected crops.[22] As the costs of these rituals were borne by the municipality, expenses and receipts for rogations are found in the accounts of ecclesiastical and civic institutions, which can give us an indication of periods of climatic stress.[23]

One of the key indicators of disaster impact (or lack thereof) and recovery is mortality, although this is sometimes challenging to reconstruct. Mortmain registers, stemming from the feudal right of the lord to part of his subjects' inheritance, and similar to the heriots used for England, have recently proved to be very useful for parts of medieval Northwest Europe.[24] Parish registers, often available there from the sixteenth century and becoming increasingly widespread elsewhere in Europe from the seventeenth century onwards, are best placed to give us an idea of mortality via burial records,[25] while baptismal and marriage records allow us to reconstruct fertility and nuptiality – variables of importance when assessing the demographic impact of crises and of possible recovery.[26] Burial records also sometimes contain direct references to diseases, and similar references can be found within ordinances, wills, theological and medical treatises, orphanage records, city accounts, and so on. Such data are patchy in early-colonial contexts, and, where

[20] See for example Wetter & Pfister, 'Spring–Summer Temperatures'; Le Roy Ladurie & Baulant, 'Grape Harvests.' See also Section 3.2.1.
[21] Walter & Schofield, 'Famine, Disease and Crisis Mortality.'
[22] Rodrigo & Barriendos, 'Reconstruction.'
[23] Piervitali & Colacino, 'Evidence of Drought.'
[24] Roosen & Curtis, 'The "Light Touch."'
[25] Curtis, 'Was Plague an Exclusively Urban Phenomenon?'; Alfani, 'Plague in Seventeenth-Century Europe.'
[26] Sella, 'Coping with Famine.'

3.1 Historical Sources

they do exist, they often relate to the colonizers rather than the colonized. However, data become much more widely available from the middle of the nineteenth century, in parallel with the rise of censuses and more formalized reports from colonies.

Demographic indicators are not the only variables affording us an idea of impact and recovery. Information on land sales, credit transactions, and criminal cases can all shed light on the severity of a crisis and on the types of coping strategies that were developed.[27] Of all types of coping strategies, perhaps those most difficult to reconstruct are informal ways of solidarity. In some regions of pre-industrial Europe, poor relief was formalized, and so accounts of poor relief institutions or overseers of the poor can offer insight here.[28] Still, more informal mechanisms were vital – even dominant in some regions – and these are much harder to investigate. Practices linked to common rights, such as gleaning, can often be traced in bylaws, but voluntary practices such as almsgiving are much harder to trace, even though there are indications that it was of huge importance.[29] Such practices are also visible in some colonial accounts, although again one must step out of the ideology and hegemonic discourse of these texts if informal coping mechanisms are to be correctly identified.

From the seventeenth and especially the eighteenth century onwards, more and more sources are at the disposal of historians. This is mainly linked to the fact that states became much stronger from the seventeenth century onwards, coupled with the growth of bureaucratic administration and colonial expansion. Alongside the rise of political economy, cameralism, and physiocracy in the eighteenth century, measurement became essential for increasing 'the wealth of nations.' This also had an effect on the types of sources linked to disasters. The Lisbon earthquake of 1755 is allegedly the first disaster in which the new focus on numbers and statistics came to the fore. The Marquis of Pombal "designed a national survey to discover the causes and origin of the natural disaster, minimize future risks and assess the damage the earthquake had caused."[30] This type of survey was also used when it came to combating disease, as for example in the Rinderpest outbreak in the late-eighteenth-century Southern Netherlands.[31]

---

[27] Campbell, 'Nature as Historical Protagonist.' For land and credit transactions, see Schofield, 'The Social Economy.' For criminality as an indicator, see Vanhaute & Lambrecht, 'Famine.'
[28] For the Low Countries, see Van Onacker & Masure, 'Unity in Diversity'; Dijkman, 'Bread for the Poor.' For England, see Hindle, *On the Parish?*
[29] Marfany, 'Quantifying the Unquantifiable?'; Lambrecht, 'The Harvest of the Poor?'
[30] Araújo, 'The Lisbon Earthquake,' 9.
[31] As mentioned in Van Roosbroeck & Sundberg, 'Culling the Herds?'

Sources originating with the government and its agencies become even more important in the nineteenth and twentieth centuries. One of the major changes throughout this period was the rapid growth in instrumental recording of environmental phenomena, as illustrated in the increase in the coverage of national meteorological networks from 1850 to 2012 in Figure 3.1.

The increased geographical coverage and temporal resolution of these data in turn allow us to pose new questions relating to societal decision-making and perceptions during hazardous events.[32] Important changes were also occurring within colonial administrations and their record-keeping: during each of the major famines in late-nineteenth-century India, for instance, the colonial authorities installed designated commissions which produced extensive and detailed reports on causes, consequences, and relief policies deemed necessary.[33] Equally, the more general annual reports compiled by colonial governments across the world provide quantitative and qualitative material with which to assemble chronologies of disasters – particularly relating to disease and famine – at regional and local scales, and also to analyze the emergence of new responses to hazards and disasters in newly colonized territories. One example is insurance, which became more and more important (linked to the emergence of a risk society).[34] Likewise, the increase of newspaper reporting (and in the twentieth century of radio and television broadcasts) that came with the expanding role of the media can yield insights on how disasters unfolded and their aftermath.[35] Media coverage has proven especially valuable for studies of the perception and 'framing' of disasters. The Lisbon earthquake of 1755 again paved the way as sensational press reports of this catastrophe, reaching audiences throughout Europe, created a novel sense of proximity and public distress over so distant an event.[36]

However, media sources, too, require intensive source critique. Photos, for instance, can depict the same disaster in very different ways. The three photos in Figure 3.2, for example, show very different sides to the Manchurian plague of 1911. They are from an album covering the plague instigated by Dr. Wu Lien-teh, a Chinese doctor who was sent to investigate the struck region and subsequently became an authority in international plague research, as well as the first president of the China Medical Association. Though all three photos are from the same album, they can nevertheless convey a disparate message regarding the situation

---

[32] Mauelshagen & Pfister, 'Vom Klima zur Gesellschaft.'
[33] Klein, 'When the Rains Failed.' [34] Rohland, 'Earthquake versus Fire.'
[35] See for example Cohn, 'Cholera Revolts.' [36] Araújo, 'The Lisbon Earthquake.'

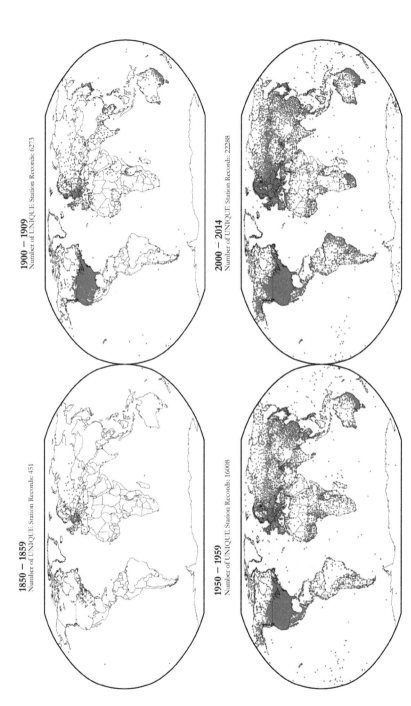

Figure 3.1 The growth of instrumental meteorological observation, 1850–2012, where each dot represents one weather station. Courtesy of Rennie *et al.*, 'The International Surface Temperature Initiative.'

Figure 3.2 Photos of the Manchurian plague of 1911, from an album instigated by Dr. Wu Lien-teh, a famous plague fighter. Courtesy of the Needham Research Institute.

3.1 Historical Sources

Figure 3.2 (cont.)

in Manchuria. The first shows white coats, white masks, and an overall 'scientific' image of matters entirely under control, legitimizing territorial jurisdiction at a tumultuous time (the final phases of the Qing dynasty). The second photo strengthens this perception of control and legitimacy further by depicting state distribution of firewood to the poor and needy. Contrarily, the third photo shows several piles of plague victims being cremated, conveying the impression that the situation was out of control, and the number of casualties uncontainable.

3.1.2 Combining Historical Data with Sources from the Natural Sciences

In recent years, historical research into disasters has grown increasingly interdisciplinary. Disaster historians have started to employ data from the natural sciences, while scholars in the natural and social sciences have begun to use historical data from the types of documentary sources introduced above, as seen most visibly in studies linking past climatic variability to human conflicts, plague outbreaks, and agricultural productivity.[37] Nevertheless, historical documents hold inherent

---

[37] See the literature discussed in van Bavel et al., 'Climate and Society.'

limitations, and, as we will see in the following section, when not viewed critically can lead to spurious conclusions.[38]

One of the most crucial limitations for the study of disasters is that the documents do not necessarily reveal the 'whole' picture: that is to say, they are not a dispassionate or objective reconstruction of the causes and consequences of a disaster. For example, we may see 'crisis situations' described in tax registers, charters, court proceedings, or colonial reports, and yet often these have the potential to be exaggerated for pleas for tax exemptions, subsidies, or charity. Equally, certain aspects of disastrous events can disappear from view in the records. Urban governments often tried to downplay the severity of an epidemic present within the city in order to maintain trading contacts and economic vitality.[39] Direct references by the urban administration of San Francisco to the earthquake of 1906 were outweighed by references to the fire that ensued. Fire prevention was simply an issue that could be more easily dealt with than earthquakes, and this reveals the selective amnesia connected to the production of documentary sources, even for a twentieth-century event.[40] Indeed, references to natural events and disasters appear in written sources only when they are relevant for the author or administration. Arable fields could experience serious erosion, but so long as they were cultivated and still provided tithes or taxes, the events would go unremarked upon. As a result, certain types of nature-induced disasters appear much more frequently in sources than others, and some types of societies over-report events, while others remain silent. In the Philippines, typhoons were registered much more accurately than earthquakes and leave accounts as far back as 414 CE, simply because they were more disruptive for humans.[41] Similarly, storm surges that did not lead to floods badly affecting human habitation or agriculture were less likely to appear in chronicles and diaries, making the distinction between cause and effect rather blurred.[42]

Non-documentary evidence can, therefore, help prevent data hiatus and help uncover source and method biases, which can allow the development of new historical interpretations of hazards and disasters of the past. Accordingly, traditional narratives on the 'late-medieval crisis' have been reinvigorated with new kinds of data in recent times.[43] Alongside the more traditionally used information on harvests, historians have recently

---

[38] Van Bavel et al., 'Climate and Society'; Roosen & Curtis, 'Dangers.' See also Section 3.1.3.
[39] Roosen & Curtis, 'The "Light Touch"'; Wilson Bowers, *Plague and Public Health*.
[40] Rohland, 'Earthquake versus Fire.' [41] Bankoff, *Cultures of Disaster*.
[42] Soens, 'Resilient Societies.'
[43] Campbell, *The Great Transition*; Pribyl, *Farming, Famine and Plague*; Bauch & Schenk, *The Crisis of the 14th Century*.

3.1 Historical Sources

begun to take note of the rapid growth in paleoclimate reconstructions derived from ice-cores, tree-rings, lake sediments, cave speleothems, and other sources.[44] This has opened up new perspectives on the Little Ice Age, its global scope, and its impacts on society.[45] The laboratory has also completely revolutionized research into plague over the past decade or so,[46] and bioarcheological evidence from skeletons in excavated burials sites is providing information on health and living standards that simply cannot be found in documents going back as far as the Middle Ages.[47]

Even better, this evidence is being actively integrated with documentary evidence. For example, the integration of paleoclimatic and written evidence of drought occurrence in sub-Saharan Africa not only provides us with a more complete view of the specific occurrence of drought itself, but also offers the opportunity to analyze why certain droughts led to human disaster while others – perhaps those of even greater severity – passed with only minor disturbance.[48] Equally, more traditional forms of archaeological research into the physical evidence – such as the increasing or declining presence of pottery shards – can provide effective comparative indicators for depopulation between regions – for example after the Black Death.[49] This is important, given that the geographical or temporal span of our documentary evidence for reconstructing the mortality effects of late-medieval epidemics is often restricted.[50] More generally, evidence from the natural sciences allows us to go back in time to periods before the widespread production of documents.

Furthermore, natural scientific data can provide an added layer of chronological development and possibly allow more accurate dating of events. For example, the dominant paradigm that disastrous sand drifts in the European coversand belt (stretching from the British Brecklands across continental Europe to Russia) increased only from the late Middle Ages onwards has been falsified by combining historical and geological data.[51] Reliance on land books, maps, and tax registers grossly exaggerated late-medieval and early-modern sand drifts, since documents of this nature in this part of Europe started to appear only from the fourteenth century onwards. As a result, earlier disasters were neglected and older dunes were dated much younger. Through new

---

[44] PAGES2k Consortium, 'A Global Multiproxy Database.'
[45] Camenisch & Rohr, 'When the Weather Turned Bad'; Hannaford & Nash, 'Climate, History, Society'; Degroot, *The Frigid Golden Age*.
[46] Little, 'Plague Historians'; Bolton & Clark, 'Looking for *Yersinia pestis*'; Green, 'Taking "Pandemic" Seriously.'
[47] DeWitte, 'The Anthropology of Plague.'
[48] Hannaford & Nash, 'Climate, History, Society.' [49] Lewis, 'Disaster Recovery.'
[50] Alfani & Murphy, 'Plague and Lethal Epidemics,' 318.
[51] Derese et al., 'A Medieval Settlement'; De Keyzer, "All We Are"; Pierik et al., 'Controls.'

techniques such as optically stimulated luminescence (OSL) dating, which dates when quartz particles became covered and were no longer exposed to sunlight, different inland dune sites can be dated more accurately. As a result, it has become clear that earlier disastrous drift sands had occurred more often than had previously been believed, and the later medieval ones relatively less often, with consequences for earlier explanatory models focusing on land reclamation or population pressure on resources.[52]

While offering great potential for the study of disasters and history, data from the natural sciences still require a critical assessment similar to that made by historians working with documentary material. The contextualization of data from the natural sciences is important for discerning the significance of the overall effect that a scientific indicator has on either a human society or a broader ecosystem. For example, in geomorphology every event of sand re-sedimentation is considered vital, with every dated layer given the same weight and importance when determining drift sand phases.[53] Yet some dunes are made up of thick layers of sand, deposited in relatively swift events, while other dunes are formed by a sequence of thin layers taking centuries to develop. It goes without saying that not all of these phases, with their varying extents and chronologies, have the same impact on human society or even on dune formation.[54] Equally, paleoclimate proxy data vary enormously in their geographical coverage and temporal resolution, so while many long-duration tree-ring chronologies of seasonal or annual resolution have been produced from the middle and high latitudes of Eurasia and North America, it is much more difficult to identify such growth increments across much of the tropics due to basic differences in climatological conditions. Natural scientific data may also have undergone statistical processing and modeling in order to make composite datasets at large geographical scales. The increasing numbers of climate reconstructions in particular regions, for example, have led to the production of what are known as climate field reconstructions. This has enabled composite time series of temperature, precipitation, and atmospheric pressure to be derived across regions, continents, and even hemispheres.[55] However, a degree of caution must be applied when using these sources to study hazards within a particular location, for they tend to be weighted to those locations in which data density is greatest (for example, Northwest Europe). Efforts to better define regional historical climate variability have also extended to Southern Africa and South America through the construction of 'multi-

---

[52] De Keyzer, "All We Are"; de Keyzer & Bateman, 'Late Holocene Landscape Instability'; Pierik *et al.*, 'Controls.'
[53] Castel, *Late Holocene Eolian Drift Sands*; Ballarini *et al.*, 'Optical Dating.'
[54] De Keyzer, "All We Are"; Pierik *et al.*, 'Controls.'
[55] Luterbacher *et al.*, 'European Seasonal and Annual Temperature.'

proxy' series.⁵⁶ These series provide valuable information for efforts to detect the nature of regional climate variability and change; however, again, the use of more localized reconstructions should be prioritized where the focus of research is on hazards within a specific area, otherwise these may be obscured or dampened by regional averaging.

These discrepancies in data availability and the uncertainty that ensues should be considered when using such data. Even more crucial when exploring the hazard–society nexus, however, is the point that similar fluctuations in environmental conditions in two different societies do not always have the same impact for humans, or even for ecosystems. Often, there is low potential for disruption to societies – and accordingly it is essential to contextualize signatures of environmental variability found within 'natural archives,' something that can be done by integrating both documentary and non-documentary sources, where available.

3.1.3   History and the Digital Age: Opportunities and Pitfalls for Historical Disaster Research

The digital age has had major impacts on the ways in which historical data have been used. Only just over a decade ago, newly constructed quantitative historical datasets such as series of prices, mortality, and disease activity were usually found in appendices of books, and the scholars who used these data – or even knew of their existence – were typically historians. Today, many datasets are either published online with the original work or digitized from an older work and hosted on publicly available online repositories, which have led to increased visibility and availability of historical data.⁵⁷ This has resulted in a wealth of opportunities for interdisciplinary scholarship into the human and environmental past. Indeed, there has been a surge of scientific interest in linking long-term human activity with environmental variability, with a new body of quantitative scholarship correlating the types of paleoclimatic data discussed in the previous section with historical data on human activity spanning the last millennium. This has led many to 'explain' human phenomena, such as conflict or disease incidence, as an outcome of climatic change.⁵⁸

---

[56] Nash et al., 'Seasonal Rainfall Variability.'
[57] An example of the former: Brecke, 'Violent Conflicts.' An example of the latter: Büntgen et al., 'Digitizing Historical Plague.'
[58] See for example Zhang et al., 'Global Climate Change'; Zhang et al., 'The Causality Analysis'; Büntgen et al., '2500 Years of European Climate Variability'; Hsiang, Burke & Miguel, 'Quantifying the Influence'; Zhang et al., 'Climate Change and War Frequency'; Tol & Wagner, 'Climate Change and Violent Conflict.'

While the very real and clear benefits of increasing accessibility of data cannot be understated, it is also important to consider what may be lost during the digitization process before historical information reaches its 'end state' as a data point that appears in a published online dataset. Such issues of source criticism are, of course, fundamental to historical research, but can go missing when old datasets are digitized and much of their contextual information is cast aside.

One such issue relates to the uneven collection and transcription of this material by scholars past and present. For example, many recently available online historical datasets are not new, but rely heavily on the work of individual scholars many decades ago. This can create spatial and temporal biases in data coverage as a result of the limited expertise and linguistic knowledge of individual scholars, or simply due to unequal access to archival material – particularly that outside of Western Europe. These problems have recently been identified in Jean-Noël Biraben's original dataset on historical plague outbreaks, according to which the Low Countries appear to have been free of plague (Figure 3.3). However, a new inventory of data collected on plague outbreaks in the Low Countries shows that this was in fact not the case, but that this gap is a legacy of the work undertaken by one particular researcher. This demonstrates the need to adopt a critical approach to what may appear to be 'complete' datasets. At worst, such issues may bring into doubt the validity and robustness of high-profile studies on the causal factors behind historical plague outbreaks, susceptibility, and spread.[59]

Digitized datasets are not limited to Europe. We have already seen the opportunities that colonial records can provide for explaining historical disasters and crises in non-Western societies, and, while such records may provide our only written sources of information at particular points in time, they must be subject to a particular type of scrutiny. One of the most frequently used historical datasets in studies linking climate and conflict incidence, for example, is Peter Brecke's global 'Conflict Catalog,' which was largely compiled from secondary published works.[60] By Brecke's own admission, this dataset is an unfinished product, with errors "especially as we go back in time and into particular regions of the world."[61] Coverage in the Southern Hemisphere – where data going back to 1400 have been used in various studies – is deficient before 1800, with most entries relating to conflicts between colonial powers and indigenous populations. Even where conflicts between indigenous populations do appear in the

---

[59] Roosen & Curtis, 'Dangers.' [60] Brecke, 'Violent Conflicts.'
[61] Brecke, 'Notes Regarding the Conflict Catalog.'

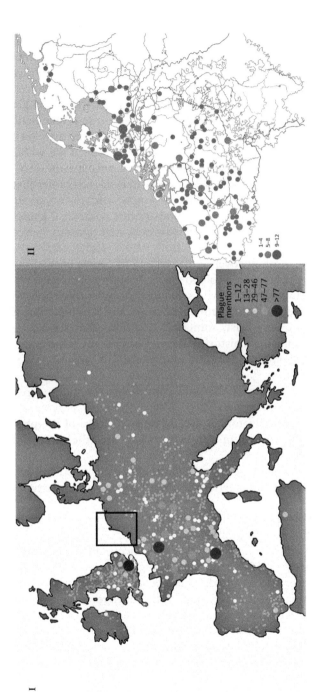

Figure 3.3 Illustration of geographical gaps in digitized Biraben plague dataset. Part (i) shows localities in Europe and North Africa reporting plague outbreaks during the period 1347–1760 according to the digitized version of the Biraben dataset. The gaps in spatial coverage are immediately visible when taking into account data for the Low Countries, indicated in the inset. When contrasted with a more recent inventory of data on locations reporting plague outbreaks in the Low Countries during the period 1349–1500 as shown in part (ii), the extent of the spatial gap for this region becomes apparent. Van Bavel et al., 'Climate and Society.'

dataset, we also find issues. The widespread conflicts of the 1820s in Southeast Africa, for example, are grouped into one decade-long conflict of the 'Zulu tribes,' a notion that dates back to early colonial writings which routinely exaggerated the effects of conflict, and in some cases even fabricated its existence.[62] The number of fatalities assigned to this conflict (60,000) is very likely based on interpretations of the same problematic sources that historians in Africa are reluctant to employ in their own studies, yet the sources on which these numbers, and indeed the whole dataset, are based are not made explicit. These criticisms are not to say that large datasets are to be discarded outside of Eurasia, but rather that new multidisciplinary efforts are needed to assess, add to, or create datasets that are based on region- and period-specific contextual knowledge, original sources rather than published works, and rigorous source critique.

As shown in recent work uncovering the history of plague in the Low Countries, historians have a major role to play in ensuring datasets are fit for purpose. This includes working with scholars from other disciplines to ensure appropriate selection, use, contextualization, and interpretation of historical data. At the very least, some key questions that should be considered before employing a historical dataset include the following.

1. Are the data geographically representative for the area(s) under consideration?
2. Is the temporal resolution of the data appropriate for the research question(s) under consideration?
3. Are the variables in the historical dataset representative of the phenomena under consideration? If not, what are the potential uncertainties?
4. Are the historical source types on which the dataset is based consistent, or do they vary? What uncertainties does variation in source types introduce?
5. How does the volume of historical source material vary over time? What uncertainties does this introduce?

Increasing specificity and transparency about uncertainty or potential biases in the data are part of the solution, though broader shifts in the publication process are also required. These could involve simple interventions like including historians as part of the peer-review teams, or more concerted efforts to develop open-access platforms through which to publish and access historical data. This would in turn incentivize historians to publish and refine datasets in a similar way to the natural sciences – the *Geoscience Data Journal* being just one example. Ultimately,

---

[62] Hannaford & Nash, 'Climate, History, Society.'

even with the most representative datasets, it is also incumbent upon scholars using these data to ensure that appropriate methodologies are selected, and it is to this issue that we turn next.

3.2 Methodologies

In this section, methodologies for historical disaster research are introduced, and the potential uses of the types of data described in the preceding section are demonstrated. This discussion of methodologies includes those employed in the reconstruction of hazards (e.g. droughts, floods, epidemics) and their impacts from historical sources, as well as analysis of human vulnerability, resilience, coping, and adaptation to these hazards. Crucially, this section stresses the importance of applying comparative methodologies over long temporal trajectories, which enables historians to move away from descriptive and event-focused approaches, although we also note how, in practice, this is not always straightforward to achieve.

3.2.1 *Hazard and Disaster Reconstruction from Historical Sources*

We have explored the types of historical documentary sources that can be used to study hazards and disasters, but how can we turn the information contained within these sources into systematic chronologies of hazard occurrence, characteristics, and impact where we lack instrumental records? Frequently the impact of hazards both past and present is couched in emotive language. An earthquake may be said to have 'decimated' Mexico City, or a tropical cyclone may be said to have 'devastated' the Mozambican coast, but to what extent can hazards and disasters be rendered comparable over space and time – and at what scales should we strive for such comparisons? In many areas, long and detailed chronologies of hazards have already been produced – particularly for source-rich regions such as Western and Central Europe. Reconstruction nevertheless remains an important part of interdisciplinary research in disaster history and its cognate subdisciplines of historical climatology and historical epidemiology. Indeed, sources previously unused for the study of disasters are still being brought to light, while improvements and innovations in reconstruction methods have allowed 'gappy' data to be used more robustly.

Documents containing abundant, regular, and systematic observations have, unsurprisingly, received most attention in the reconstruction of past hazards. One of the major reasons for this is that sources of this nature usually lend themselves to the application of statistical methods, especially where there is an overlap in observations of the same variable between historical documents and modern data. In turn, this allows for the calibration of statistical relationships between the recorded variable (e.g. wind direction) and the hazard to which it relates (e.g. precipitation), which can then be applied to historical data.[63] This principle can apply to a range of historical sources, with perhaps the most famous example being harvest dates and temperature.[64] In some regions, the inverse approach has been adopted, where paleoclimatic data on temperature have been used to reconstruct agricultural yields.[65]

In large parts of the Southern Hemisphere, the largely qualitative nature of historical documentary sources means that statistical methods are less suitable and other methods such as textual and content analysis need to be used. Typically, this means that a body of hazard-related quotations within a season or year is assessed against a quantitative scale. In the case of rainfall reconstruction from missionary accounts and private diaries, for example, such a scale might range from drought (−2), dry (−1), 'normal' (0), wet (+1) to extremely wet (+2) relative to 'typical' rainfall levels.[66] The resultant seasonal or annual values can then be combined to produce long-run time series of semi-quantitative climate data, which can be calibrated for accuracy in any overlap that exists with early instrumental series. Similar methods have been used to reconstruct cyclones, whereby qualitative descriptions of atmospheric conditions and of the extent of damage to buildings have been used as a proxy for cyclone intensity.[67] Even where observations on hazards are too sparse or scattered to reconstruct seasonal or annual variability in this way, however, they should not go to waste. We have already seen how historical observers tended to record those hazards which led to some form of societal impact. These subjective descriptions of climate-related hazards may be as important for the historian concerned with disasters as an 'objective' instrumental record, since disasters themselves are the product of human as much as environmental factors and are often subjectively defined phenomena.[68]

Moving beyond the reconstruction of hazards themselves, three main categories are accepted to measure the impact of a disaster: effects on people (death, injury, disease, and stress), effects on goods (property damage and economic loss), and effects on the environment (loss of flora and fauna, pollution, and loss of amenity). We will focus on the

---

[63] Hannaford, Jones & Bigg, 'Early-Nineteenth-Century Southern African Precipitation.'
[64] Pribyl, *Farming, Famine and Plague.*
[65] Huhtamaa & Helama, 'Reconstructing Crop Yield Variability.'
[66] Nash et al., 'Seasonal Rainfall Variability.'   [67] Nash et al., 'Tropical Cyclone Activity.'
[68] Hannaford, 'Long-Term Drivers'; Miller, 'The Significance of Drought.'

3.2 Methodologies

use of mortality as an indicator in the following section. Mortality is by far the most dominant category employed to measure the impact of a disaster, and it gives us a clear insight into the challenges when it comes to measuring impact. Mortality is commonly at the center of thresholds and levels of minimal disruption used to identify and classify disasters.[69] For example, the Centre for Research on the Epidemiology of Disasters (University of Louvain), which hosts one of the most extensive modern disaster databases, includes only those events with more than 100 deaths, as well as damages amounting to 1 percent or more of GDP, and the number of people affected as 1 percent of the total population.[70]

This type of assessment is, however, very much focused on physical damage and mortality, and excludes other important facets of disasters that we have discussed in Section 2.1. Another strategy is to define a perimeter that is hazard-dependent but uses a fixed threshold. Mark and Catherine Casson defined a crisis in terms of a deviation from an attested mean value – for example, rising mortality rates and soaring average food prices that progressed 20 percent beyond the normal average were deemed a crisis.[71] This can, however, run into the same problems highlighted above, in that fixed values and thresholds cannot be applied to all hazards and are to some extent context-dependent.[72] Other studies draw more directly from the concept of ecological resilience, defined as "the magnitude of disturbance that can be absorbed before the system redefines its structure by changing the variables and processes that control behavior."[73] Often this is applied to ecosystems, but it has been used for societies and social institutions as well. With this definition, an event can be labeled a disaster when it redefines the structure and behavior of a society.

Mortality is an indicator of particular importance when reconstructing epidemic outbreaks – a trend that has been reignited in recent years with moves towards digitalization of data and GIS mapping. The sources necessary for reconstructing epidemics can be divided into two broad groups: direct documentary references or mentions of a disease that can be used as one form of 'diagnostic' evidence, or epidemiological evidence referring to the severity or spread of a disease, in its entirety or – preferably – differentiated by sex, age, and socio-economic status, as well as separating rural and urban data.

---

[69] See Section 2.2.   [70] Smith, Keith & Petley, *Environmental Hazards*.
[71] Casson & Casson, 'Economic Crises in England.'
[72] A point we will return to in Section 3.2.3.
[73] Gunderson, 'Ecological Resilience,' 426. See Section 2.3.4 for a detailed breakdown of the concept of resilience.

Direct references to diseases can be found in various pre-modern sources, including burial registers, ordinances, wills, theological and medical treatises, bills of mortality, orphanage records, and city accounts. While these records can provide useful pointers to the dominant epidemiological conditions of a certain season or year, epidemiological information on a disease within such documents is not direct diagnostic information on what that disease actually was, and we should not always take the diagnosis of pre-modern contemporaries at face value.[74] The term '*peste*,' for example, may have referred to plague, but also may have been a 'catch-all' term to refer to all kinds of different afflictions that may have had features similar to plague.[75] In Western Europe, it is often difficult to find a systematic distinction between diseases prior to the second half of the fifteenth century – though in the early-modern sources this becomes much clearer. For example, scholars have been rather damning of what the burial records can directly say about cause of death,[76] but in more recent times certain burial records have been shown to make very sharp distinctions between diseases – often with numerous terms used even in the same year.[77] Nevertheless, even if pre-modern scholars, especially by the sixteenth and seventeenth centuries, could make broad distinctions between diseases,[78] we still must accept that their diagnoses could be wrong – only DNA pathogen evidence can ultimately solve this. For some parts of the world, this lack of absolute confirmative evidence for what a disease actually was has led to substantial disputes: some scholars suggesting, for example, that plague did not substantially feature in pre-modern Japan,[79] and it has been questioned whether the Chinese term '*wenyi*,' broadly translated as plague, was actually the same disease caused by *Yersinia pestis*,[80] even if we now have strong evidence that suggests the initial outbreak of the Black Death was linked to strains of the *Yersinia pestis* pathogen originating in the Qinghai Plateau of Western–Central China or southern Siberia.[81]

Mortality data also offer the opportunity to quantify the occurrence, spread, and severity of epidemics. In recent years there has been an attempt to quantify Black Death mortality rates across numerous localities within Europe.[82] However, the source material behind this exercise is

---

[74] Cohn, *Cultures of Plague*; Carmichael, 'Universal and Particular.'
[75] Theilmann & Cate, 'A Plague of Plagues.'
[76] Dobson, *Contours of Death and Disease*, 81.
[77] Curtis, 'Was Plague an Exclusively Urban Phenomenon?'
[78] As suggested in Cohn, *Cultures of Plague*.
[79] Bowman Jannetta, *Epidemics and Mortality*.
[80] Dunstan, 'The Late Ming Epidemics'; Cao & Li, *Shuyi*.
[81] See the classic Cui *et al.*, 'Historical Variations.'
[82] Christakos *et al.*, *Interdisciplinary Public Health Reasoning*; Voigtländer & Voth, 'The Three Horsemen'; Gómez & Verdú, 'Network Theory.'

3.2 Methodologies 65

invariably not epidemiological. It consists of estimates of mortality impact by contemporary observers from all kinds of disparate sources, which, in truth, are difficult to compare. More reliable epidemiological indicators come from sources which have roughly consistent features between localities and even over time, and, of course, exist for a large number of localities – mortmain accounts and church burial records being two of the main examples, although one problem has been that material for epidemiological reconstructions over large areas and long periods is not as readily available for the late Middle Ages.[83] Methodologically, obtaining mortality rates from these data sources is problematic as these require either recorded population estimates or highly sophisticated numerical models in order to produce the information. For England, the latter have been developed by the Cambridge Group for the History of Population and Social Structure, and are expounded at length in the *Population History of England*.[84] Other methods beyond simply listing the total numbers of burials have also been used. An example that has regained popularity is one originally developed by the Italian demographer Massimo Livi Bacci, whereby relative annual mortality is calculated on the basis of the percentage increase or decrease in burials from preceding years, offering insight into the severity and spread of mortality across different localities and regions.[85] Because an increase in burials to a level 50 percent or more higher than in previous years has been suggested to have prevented the generation born in a given year from replenishing the population, this threshold has been used as a sign of crisis or disaster.[86] Two of the major advantages of this method are that it requires no data other than the burial records themselves, and is amenable to relatively straightforward calculations and processing. Analysis of seasonal distributions of burials can also give insight into causes of death, which may include famine as well as disease. Age and gender may offer further clues, although, given the uncertainties around the age and gender characteristics of some diseases, one should be careful to avoid circular reasoning here.

3.2.2 Vulnerability Assessment

A vulnerability assessment analyses how exposed certain individuals or groups are to a particular hazard. It tracks the potential population at risk and tries to explain the social structures, economic behaviors,

---

[83] For a discussion of these sources, see Section 3.1.1.
[84] Wrigley & Schofield, *The Population History*.
[85] Roosen & Curtis, 'The "Light Touch"'; Alfani, 'Plague in Seventeenth-Century Europe'; Curtis, 'Was Plague an Exclusively Urban Phenomenon?'; Curtis & Dijkman, 'The Escape from Famine.'
[86] Alfani, 'Plague in Seventeenth-Century Europe.'

institutional conditions, and physical circumstances that determine the exposure to hazards and the ability to recover from their occurrence. Sometimes vulnerability is identified simply through the occurrence of disaster, but this is problematic: at least in theory, vulnerability may exist irrespective of whether a disaster occurs, even if in practice the underlying patterns may be revealed only when disaster does strike.

One way to unravel – although not formally assess – vulnerability is suggested by the Pressure and Release (PAR) model developed in disaster studies.[87] This model aims to explain how the interaction of vulnerability, on the one hand, and the occurrence of hazards, on the other, may lead to disaster. In the process, it also unpacks the concept of vulnerability. The 'pressure' component of the model distinguishes three levels of underlying factors. At the base are 'root causes': economic, social, and political conditions that affect the distribution of power and resources. 'Dynamic pressures,' such as rapid urbanization, economic depression, or war, may transform these root causes into the 'unsafe conditions,' such as settlement in hazard-prone locations or unhealthy living quarters, that are the ultimate expression of vulnerability. The concepts of root causes, dynamic pressures, and unsafe conditions have also been used as a lens through which vulnerability in case studies of historical disasters can be explained.[88]

A more formalized assessment of vulnerability has also gained currency within contemporary fields such as development or climate studies. This typically draws upon quantitative social research methods such as household surveys, or may involve the analysis of larger economic datasets. A common approach is the use of indicators. This involves the identification of a series of indicators associated with exposure, sensitivity, and adaptive capacity to a particular hazard, independently of the occurrence of the hazard itself.[89] These indicators can be based upon links established from a range of general and context-specific literature, for example that diversity in cultivated crops reduces sensitivity to drought, or that grain storage enhances adaptive capacity. Other indicators may be more subjective, where, for example, the integration of pre-colonial African communities within intercontinental trade networks on the one hand provided a potential source of grain in times of scarcity, but on the other hand subjected these communities to exploitation. Indicators are usually clustered in a limited number of relevant dimensions and subsequently translated into an index. Typically, indicators are given a value that represents a positive or negative contribution to vulnerability. These

---

[87] Blaikie et al., *At Risk*, 24.
[88] Barnes, 'Social Vulnerability and Pneumonic Plague'; Soens, 'Resilient Societies.'
[89] Hinkel, "Indicators of Vulnerability"; Füssel, 'Vulnerability'; Hahn, Riederer & Foster, 'The Livelihood Vulnerability Index.'

3.2 Methodologies

values may be weighted according to those components adjudged to contribute the most to vulnerability, or they may simply be given equal weight. The result can be a snapshot of vulnerability at a particular point in time for a particular locality or social group, but it is also possible to make temporal or spatial comparisons by including multiple localities, regions, or time periods in the analysis.

Although in historical research assessment of vulnerability is usually less formalized and frequently qualitative in nature, indicator analyses have recently been applied to various historical contexts, notably the Irish famine of 1740–41,[90] climate anomalies in Iceland and the Eastern United States during the first and second millennia AD,[91] and drought in Southeast Africa between the sixteenth and nineteenth centuries.[92] This range of contexts demonstrates the particular value of indicator approaches in areas where sources – which may also be archaeological or zooarchaeological – are fewer in number but are of sufficient breadth to capture the multidimensional nature of vulnerability. Where documentary sources are greater in number, the information available on one indicator (e.g. wages) may go far beyond that available for a whole cross-section of indicators in other contexts, and so the use of indicators may be a less suitable approach. There is also a difficulty in integrating cultural norms and religious beliefs into indicator assessments, which can render them somewhat two-dimensional. Thus, while indicator approaches provide a useful comparative tool, the results should generally be seen as points of departure that seek to simplify a complex reality rather than end-states of analysis.

Whichever way vulnerability is analyzed, in historical perspective we can find merit in considering vulnerability as the 'flipside' of adaptation – that is to say, vulnerability and adaptation are co-evolving, interdependent phenomena.[93] Bringing adaptation into the discussion also calls for an analysis of institutional responses and the social actors behind these responses. We have already seen how responses to hazards and disasters are not necessarily 'rational' or equivalent to the 'common good,' and this is where qualitative analysis of institutional records of a narrative nature can provide an invaluable counterpart to analysis of both pre-existing vulnerabilities and hazard and disaster outcomes. This allows us to bring into view the importance of contingency, social actors, and environmental knowledge in human–environment interactions over time, while also helping us avoid linear or teleological success narratives of adaptation as 'improvement.'

---

[90] Engler et al., 'The Irish Famine.' [91] Nelson et al., 'Climate Challenges.'
[92] Hannaford, 'Long-Term Drivers.'
[93] See arguments made in Rohland, 'Adapting to Hurricanes.'

3.2.3 Comparative Methodologies

This chapter started out by pointing to the importance of comparative analysis. Before we expand on how this can be done, we must first consider why it is necessary and the problems associated with some of the other approaches that have been prominent in historical research on disasters over recent decades. These approaches are linked to the move that the historical profession has made away from the social sciences, resulting in the situation where historians are less likely to subscribe to the view that the past can be used to establish regularities, patterns, and certainly not laws, through comparative analysis. This has been driven in part by a fear of accusations of being 'deterministic,' therefore privileging events and the narrative, and perhaps even a situation where post-structuralists would offer up the past as an "undecidable infinity of possible truths."[94] For the specific study of historical disasters, there are a number of drawbacks to this present situation of the historical profession.

First, a focus on events, as in *histoire événementielle* or evental history, can lead to over-emphasis or over-exaggeration of certain features within these 'special cases' that are then said to apply more broadly for other hazards and the disasters that can ensue. For example, the notion that epidemics inevitably caused societies to descend into scapegoating and persecution of easily targeted groups has undoubtedly been connected to scholars focusing in on just a few very spectacular cases such as the Black Death or AIDS – which have proven to be anomalies when placed within a broader temporal and geographical perspective of all social responses to epidemics.[95]

Second, over-emphasis on one disaster can lead to the privileging of certain theoretical or explanatory frameworks over others. For example, for the medieval period, an exceptional amount of focus has gone into the famine of 1315–17 in isolation, which is problematic in the sense that this famine occurred – at least in a European context – within very special social and environmental conditions that were probably not to be repeated across the whole of the pre-industrial period to the same extreme. That is to say, the famine of 1315–17 in many parts of Northern Europe occurred in conditions of unparalleled population pressures on resources – thus being an exceptional event – but not one representative of all the famines that occurred throughout the medieval and early-modern periods.[96] As yet, few attempts have been made to compare, for example, the famine experience of localities in conditions of high Malthusian

---

[94] All discussed in Curtis, van Bavel & Soens, 'History and the Social Sciences.'
[95] Cohn, *Epidemics*.   [96] A problem highlighted in Geens, 'The Great Famine.'

3.2 Methodologies

pressures with the famine experience of localities with low Malthusian pressures. Similar concerns over the representativeness of historical hazards chosen by scholars have recently been iterated by historians and archaeologists focusing on late-medieval earthquakes in the Mediterranean.[97]

Third, this can also lead to the problem of over-emphasizing a conspicuous feature of a society hit by a disaster as possible causal factor. Quite often, at least in studies focusing on the twentieth and twenty-first centuries, the factor that is often cited is 'poverty' by way of description or inductive reasoning – and this feeds into narratives that see the 'Global South' as a disease-ridden, inhospitable place – poverty-stricken and disaster-prone in equal measure.[98]

One of the positive developments in recent years in the study of historical disasters is the greater tendency towards assessing hazards and shocks, and the disasters that can ensue, in a much broader geographical and temporal perspective – that is to say, we are moving in a more global direction. Recent literature has warned us of over-focusing on the Black Death, and trying to apply abstract theoretical models of redistribution or economic development based on the logical mechanisms connected to this 'anomalous' shock.[99] Scholars are now assessing the Black Death within its place in the broader chronology of the Second Pandemic over five centuries.[100] In line with broader movements in the discipline of history, scholars have also moved in more recent times to challenge Eurocentric conceptions of the Black Death, instead using new kinds of global data sources – particularly from bioarcheology and genome analysis – to reveal its truly global effects and consequences.[101] On the subject of famines, recent literature too has moved away from looking only at very severe cases such as the 1315–17 Great Famine. Instead, it considers how food crises and famines developed in regions across the whole pre-modern period – comparing some of their features, causes, and consequences in a more standardized way.[102] Furthermore, we have come to the realization that a terrible event etched in popular consciousness such as the Great Famine of Ireland in the middle of the nineteenth century was not a catastrophe for Ireland in isolation, but part of a broader problem affecting much wider geographical territories – and

---

[97] Forlin, Gerrard & Petley, 'Exploring Representativeness and Reliability.'
[98] Frerks & Bender, 'Conclusion,' 199.
[99] Alfani & Murphy, 'Plague and Lethal Epidemics.'   [100] Cohn, 'Patterns of Plague.'
[101] The special issue of *The Medieval Globe* edited by Monica Green, *Pandemic Disease*, https://scholarworks.wmich.edu/tmg/vol1/iss1/. Also see Green, 'Black as Death.'
[102] Alfani & Ó Gráda (eds.), *Famine*; Collet & Schuh (eds.), *Famines during the 'Little Ice Age.'*

that the excessive focus on events such as those in Ireland is entrenched in historiographical traditions.[103] Scholars are also now interested in placing earthquakes in broader geographical and temporal frameworks. For a long time earthquakes were the subject of analysis as isolated events, and most comparative approaches were limited to those interested in paleoseismological aspects of earthquakes – resulting in mere catalogs of seismic events.[104] Recent research has looked to add to that approach by elaborating upon similarities and differences between Mediterranean late-medieval and early-modern societies in coping with and preparing for earthquakes.[105]

There are, however, practical challenges to employing comparative approaches – which are applicable for the discipline of history as a whole, but with even greater relevance for the study of historical disasters. As we have seen in the previous sections of this chapter, the first issue is that historical source material is a fickle substance, distributed unevenly across time and space. Even when sources are relatively abundant, they are often difficult to interpret and reveal only small segments of the phenomenon under investigation.[106] This can be demonstrated in the study of the redistributive effects of disasters for dimensions such as wealth and property. Each source for reconstructing inequality has its own idiosyncrasies – differing methods of calculation, recording, and exclusion rates – which mean that it is difficult to compare the redistributive effects of disasters between regions or localities in any absolute terms. Some scholars have conceded this and suggested that these kinds of sources can be used only in very relative terms – measuring the redistributive effects of shocks such as epidemics or floods in the same localities or regions over time using the very same type of source.[107] Indeed, we have come to realize that it is in the temporal dimensions of comparison that trained historians have considerable advantages and can provide added value when trying to understand the causes and consequences of disasters. Another source-related problem connected to comparative approaches is the simple issue of their complete absence – or rather random appearance over time. In keeping with the same example of redistribution, the problems can be seen in Walter Scheidel's *The Great Leveler* on the redistributive impact of terrible mortality shocks and

---

[103] See the essays in Ó Gráda, Paping & Vanhaute (eds.), *When the Potato Failed*.
[104] For a selection of online catalogues see www.bgr.bund.de/EN/Themen/Seismologie/E rdbebenauswertung_en/Kataloge_en/historisch/historische_erdbeben_inhalt_en.html
[105] Forlin & Gerrard, 'The Archaeology of Earthquakes.'
[106] Curtis & Roosen, 'On the Importance of History.'
[107] Alfani & Ammannati, 'Long-Term Trends in Economic Inequality'; van Bavel, Curtis & Soens, 'Economic Inequality.'

3.2 Methodologies

outbreaks of violence.[108] The only empirical study cited in Scheidel's book that actually has this very specific kind of systematic information for the pre-industrial period (immediately before and after a shock) concerns one town in Northern Italy in the seventeenth century.[109] Put simply, throughout history, we have very few occasions where we have an empirically measurable record for distribution just before a disaster and then just after a disaster, and it is even rarer to have both of those (which would be necessary for assessing direct short-run effects) and then a long-term series of the same information. This reduces our confidence in the posited causal links between, for example, 'leveling' and catastrophic shocks. And this does not merely apply to redistribution but to a whole raft of social indicators – the same problems can be posited for the effects of disasters on age of marriage, for example.[110]

A second problem explicitly related to the subject of historical disasters is that it is very difficult to apply systematic social science methods: an obvious one being to hold a number of variables constant as much as possible, in order to isolate those that are held to be crucial and are tested for. Although the geographical and temporal scope of our comparisons has broadened with regard to historical disaster studies,[111] systematic comparative approaches lag some way behind. In his discussion of social science concepts and comparative methods, Giovanni Sartori noted that "If two entities are similar in everything, in all their characteristics, then they are the same entity. If, on the other hand, two entities are different in every respect, then their comparison is nonsensical."[112] Disaster studies scholars have to work more on reducing the number of variables – one of the most important components of successful comparative research – if they are going to get closer to answering important social science questions such as why some societies cope more effectively than others with hazards. The problem here is that quite often we are comparing different hazards, or at least different magnitudes and intensity of the same type of hazard, at the very same time as comparing different societal variables of possible importance.[113]

This issue can be demonstrated by focusing on just one typical kind of failure in this regard: for example, in the comparison undertaken between two floods occurring in 1993 and 1994 in two different

---

[108] Scheidel, *The Great Leveler.*
[109] Investigated by Alfani, 'The Effects of Plague.' See also Curtis, 'All Equal in the Presence of Death?'
[110] See Section 6.1.1.   [111] See the recent Schenk (ed.), *Historical Disaster Experiences.*
[112] Sartori, 'Comparing and Miscomparing,' 246.
[113] As argued at length in van Bavel & Curtis, 'Better Understanding Disasters.'

countries – in Northwest Italy and the US Midwest.[114] In conclusion the author suggests that very divergent 'human responses' to the floods were connected to essential differences between the afflicted societies in terms of socio-political traditions and organization and levels of integration within communities. This may, of course, prove to be correct, but the argumentation is overshadowed by the fact that we are never sure what impact the differing magnitude and scale of the respective floods had on societal responses. It thus becomes difficult to separate factors at the local or national level, and it is unclear what differences exactly played a crucial role in light of the numerous differences between these two societies. What could have been more illuminating is to take the research one stage further and compare responses of different localities *within* either the US Midwest or Northwest Italy – limiting the number of possible independent variables and thus being more able to hold the hazard constant. In fact, this is where the pre-industrial period actually has some advantages over these kinds of modern studies because, on a regional and even local level, small-scale societies which were very close to each other (separated by just tens of kilometers) could have very divergent economic and agricultural organization, microdemographic regimes, and patterns of tenure and resource distribution, among other factors.[115] Very different societies close to each other and exposed to the same exogenous pressure can be identified – a good example being the regional comparison of different plantation economies dealing with the eruption of Mount Soufrière in 1812 on the island of St. Vincent, and actually showing very divergent rates of recovery.[116] Furthermore, this is generally something far easier to find in the historical context than for the twenty-first century – particularly in the 'developed' world, where these differences have become less sharp under the homogenizing forces of modern commerce, the rise of nation states, improved transport, and better communication.

Smaller-scale comparisons over longer periods of time offer a way forward for comparative approaches into historical disasters, and that in itself may bring with it new opportunities for other kinds of comparisons – often within the same individual localities or communities themselves. This can be across many dimensions which are often underexplored in the historical context – the differential impact of a disaster occurring within the same locality over time between men and women, adults and children,

---

[114] Marincioni, 'A Cross-cultural Analysis.'
[115] Curtis, *Coping with Crisis*; van Bavel, *Manors and Markets*.
[116] Smith, 'Volcanic Hazard in a Slave Society.'

the healthy and the frail, the rich and the poor, and recent migrants and (so-called) natives. Smaller-scale comparisons may also avert some of the problems relating to disparities in source types and availability that come with larger-scale comparisons. There is still a great scope for analytical, smaller-scale comparative work on historical disasters.

4 Disaster Preconditions and Pressures

This chapter looks at the preconditions and pre-existing pressures that help determine the impact of hazards, and the potential disasters that can ensue. Although many disaster studies look at the effects that unfold after the occurrence of a disaster, preconditions and structures that were present before a shock are as potentially important for explaining why a hazard can turn into a disaster as the immediate and long-term responses. Hazards take place within environmental and social contexts that shape or even determine how a disaster unfolds, and how a society or social groups can respond. These might be considered the core components behind the resilience of societies and vulnerability of different social groups to hazards.[1] Some of these pressures and preconditions develop slowly and incrementally or are just basic features of a particular region, while others are short-term pressures that arose just before the occurrence of a hazard such as warfare, rebellions, or migration. In this chapter, we distinguish between a number of different pre-existing pressures: climatic and environmental conditions, levels of technology, and the state of economic development, but also pre-existing pressures connected to social organization. This includes institutional configurations and societal coordination systems, levels of poverty and inequality, and cultural aspects. Nevertheless, it must also be noted that while we consider these as potential preconditions to a disaster, none of these pressures occur out of nothing. Indeed, as the disaster cycle framework implies,[2] they are frequently in some way the result of an earlier hazard risk or disaster.

4.1 Environmental and Climatic Pressures

A prime factor, and an intrinsic aspect of risk, is geography.[3] Every place on earth has a particular geological, environmental, and climatic setting which defines the underlying hazard exposure of that specific region and thus contributes to levels of vulnerability. Earthquakes occur along fault

---

[1] See Section 2.3.3. [2] See Section 2.3.2. [3] Hewitt, *Regions of Risk*, 12.

4.1 Environmental and Climatic Pressures

lines in the earth's surface, droughts are a frequent life experience in the Sahel, the coastline between Bordeaux and Schleswig-Holstein is prone to flooding, while malaria is limited to the habitat able to support sufficient quantities of the Anopheles mosquito. This is called biophysical vulnerability. Every region on earth struggles with at least one, but possibly multiple, biophysical vulnerabilities.

The 1930s Dust Bowl in the American Great Plains, for example, can only be understood by taking the dual biophysical vulnerability of that region into account: droughts and aeolian soils. Droughts have been a long-standing hazard in the Great Plains, as the informal name Great American Desert suggests. Simulations have shown that the Great Plains witnessed periods of drought, such as the one in the 1930s, at least four times between 1900 and 1950. While the severity of the 1930s drought was perhaps exceptional, major droughts have occurred in the Great Plains once or twice a century over the past 400 years.[4] The cause of these droughts can be attributed to changing sea surface temperatures, with a strong correlation detected between varying Pacific sea surface temperatures and periods of low precipitation in the Great Plains of America.[5]

Droughts, however, do not cause major erosion and sand drifts unless they are combined with fine-grained soils from wind-deposited sediments, such as those that characterize the Great Plains. These aeolian soils have been deposited throughout the Holocene, and are prone to sand drifts and erosion when not covered by sturdy vegetation. For a long time, the region had been considered unsuitable for agriculture and was covered by prairie vegetation. In the late nineteenth and early twentieth century, however, the search for productive and commercial land changed the landscape into one dominated by cattle ranches and later arable land. The soil was plowed and laid bare, heightening biophysical vulnerability by exposing the inherently erosion-prone soils to the winds that swept across the American plains.[6] The Great Plains of America can be seen, therefore, as an example of a 'region of risk.' The concept was coined by Kenneth Hewitt and defines a geographical region that is characterized by recurrent natural hazards of a certain type.[7]

Vulnerability resulting from environmental conditions figures prominently in policy and risk evaluation reports. Indeed, as noted by Bankoff, the idea that "disasters are simply unavoidable extreme physical events that require purely technocratic solutions still remains the dominant

---

[4] Schubert et al., 'On the Cause of the 1930s Dust Bowl,' 1858.
[5] Schubert et al., 'On the Cause of the 1930s Dust Bowl,' 1855.
[6] Lee & Gill, 'Multiple Causes of Wind Erosion.'   [7] Hewitt, *Regions of Risk.*

paradigm within the UN and multilateral funding agencies such as the World Bank."[8] As a result, no efforts are spared to map these risks and vulnerabilities. The British government, for example, has created and shared an interactive map showing flood risks in all of England.[9] Similar projects and risk assessments exist for every type of biophysical and chemical risk. In Northwest Europe, floods along the North Sea coast have been a frequent life experience, and coastal communities face this constant biophysical vulnerability. In the struggle against floods there from the Middle Ages to the present, risk analyses have plotted possible winter storms and flood hazards. Certain regions are considered much more hazard-prone than others, however. Western countries have helped establish a discourse that distinguishes between themselves as safer regions and the rest of the world that is inherently considered more risky or unsafe.[10] The edge of the Pacific Ocean, for example, stretching from Australia to East Asia and the American West Coast, is often called the 'Ring of Fire,' because of the subduction of tectonic plates, leading to high frequencies of earthquakes and volcanic eruptions. Similarly, tropic zones are consistently marked as dangerous, with warnings concerning infectious diseases and health risks. Societies living in hazard-prone zones of this type are often considered highly vulnerable, regardless of the prevention measures and mitigation strategies implemented to cope with these hazards.

Biophysical vulnerability has often been considered as forming a static backdrop to human affairs, with the geological and environmental characteristics of particular regions seen as a given set of circumstances that will continue to affect societies throughout time. However, increased knowledge of changing climatic conditions past and present is changing this idea fundamentally. In the fields of historical climatology and paleoclimatology, progress has been made in mapping changing biophysical conditions in the past. For example, the Sahel has not been affected by the same drought conditions throughout the Holocene. Owing to climatic shifts, the biophysical hazard of drought in this particular region of risk has experienced multiple changes. The shift from the Medieval Warm Optimum to the Little Ice Age increased humidity in this region, with obvious implications for susceptibility to drought.[11]

Nevertheless, even 'global' climatic shifts like the Little Ice Age mapped out onto particular regions in different ways – especially through

[8] Bankoff, 'Rendering the World Unsafe,' 25.
[9] Long term-flood risk information, https://flood-warning-information.service.gov.uk/long-term-flood-risk/map (last visited 18 April 2019).
[10] Bankoff, 'Rendering the World Unsafe.'
[11] Carré et al., 'Modern Drought Conditions', 1949.

the changes in precipitation levels that sometimes accompanied cooler temperatures. Sam White has convincingly shown that in the Middle East, the Little Ice Age did not bring the same type of weather patterns that Western Europe encountered. The North Atlantic Oscillation that influences Northwest Europe as well as the Middle East had an opposite impact in the two regions: while Northwest Europe was confronted with wetter conditions, the Middle East experienced increased droughts and more frequent extreme cold spells.[12] Equally, societies in Southeast Africa experienced generally drier conditions throughout much of the Little Ice Age, while those in Southwestern Africa were confronted with wetter conditions.[13]

Biophysical vulnerabilities therefore can and do change throughout history. This realization has become even more important given current climatic change, with extreme climatic events as well as longer-term changes at least partly linked to human agency and responsibility. It is important to map these temporal and geographical patterns to fully understand regions of risk and their inherent biophysical vulnerabilities. Notwithstanding the importance of shifting environmental baselines, one of the underlying premises of this book is the social vulnerability approach: risks do not simply arise from environmental circumstances but are almost always shaped by social vulnerabilities and adaptability. The next sections therefore explore societal preconditions of hazards and disasters.

4.2 Technological, Infrastructural, and Economic Preconditions

4.2.1 Technological and Infrastructural Preconditions and Pressures

Technological and infrastructural changes are often linked to mitigation measures and long-term changes after the occurrence of a disaster, and we address those changes and effects in Chapter 5. Equally important, however, are the technology and infrastructure already present before a hazard or shock, which help us to understand how a disaster unfolds.

Societies develop their technology and infrastructure under particular economic, social, and political conditions. Agricultural technology in sub-Saharan Africa can provide an illuminating example. One of the factors contributing to the persistence of famine in Africa up to the present day is low agricultural productivity. In the 1950s and 1960s, agriculture in the Western world, but also in large parts of Latin

---

[12] White, *The Climate of Rebellion*.   [13] Hannaford & Nash, 'Climate, History, Society.'

America and Asia, was transformed by technological innovations such as high-yielding or drought-resistant varieties of cereals, chemical fertilizers, pesticides, and improved techniques for irrigation, which allowed agricultural production to keep up with population growth or even exceed it. Large parts of Africa, however, lagged behind: an African 'Green Revolution' did not materialize. Even today, many African smallholders have only very limited access to the technological means to raise productivity. Yields are very low: in fact, food production per capita has declined by about 10 percent since the early 1960s.[14] Obviously, behind this chronic under-production lie other factors: widespread poverty, limited access to education, and a lack of government support for subsistence agriculture, for example. Furthermore, factors not related to agricultural production also contribute to the persistence of famine, such as weak internal markets, armed conflict, poorly functioning governments, and perhaps also the liberalization of the global food trade. The fact remains, however, that Africa's vulnerability to famine is in part determined by the state of technological development of its agriculture.

In many cases infrastructure and technology were shaped not just by general economic, social, and political conditions, but also by responses to previous hazards and shocks. Until the nineteenth century, most inhabitants of European and North American cities were dependent on water pumps and wells for their drinking water. The introduction of piped-in water supply systems was partly a reaction to the health risks posed by contaminated drinking water. The availability of running water, however, led to a large-scale increase of water consumption. Among the many innovations introduced were water closets in the houses of the well-to-do. These water closets were not yet connected to a sewage system, but used underground vaults and cesspools as the main manner of disposal. Leaking vaults and overflowing pools could easily infiltrate the water tables used for the remaining wells and pumps, actually increasing health hazards instead of reducing them. Contaminated and polluted water, therefore, became an increasing pressure in changing cities.

This combination of risks could eventually lead to disaster, as happened in London, when the bacterium *Vibrio cholerae* was found in a water pump in Broad Street, causing an epidemic outbreak of cholera. By carefully mapping the first casualties, physician John Snow identified the infected pump and concluded that cholera was spread via contaminated water in 1854. Accordingly, infrastructure and available technology in the populous and crowded nineteenth-century cities had created the

---

[14] Devereux, 'Why Does Famine Persist in Africa?'; Ó Gráda, *Famine: A Short History*, 263–266.

preconditions for the recurrent outbreaks of infectious diseases that characterized most of that century.[15] These outbreaks of cholera alarmed governments all around the Western world and triggered investment in large-scale public works, such as the development of sewage and water systems. As Tarr noticed, however, most of the technological solutions for these pressures, such as new sewage and water systems, created new risks and pressures of their own. Running-water sewages, for instance, created water contamination in rivers and streams, negatively affecting settlements downstream – places previously unaffected.[16] Again the technological jump forward created new pressures for regions and cities previously not affected. This build-up of risks shows how infrastructural or technological changes are never implemented on a 'clean slate' and that preconditions are hardly ever entirely independent from former hazards and shocks.

An even clearer example of the interplay between mitigation measures and preconditions is provided by coastal infrastructures along the North Sea shoreline. In the early and high Middle Ages, most communities in this region developed tidal economies. Instead of tackling floods and regulating tidal flows, these societies found ways to co-exist with a tidal landscape and adapt to the marshy and wet conditions. Fishing, salt production, and grazing, using the resources provided by the salt marshes, became the dominant economic activities along the North Sea shores. All these activities benefited from the recurrent seasonal inundations and were unthreatened by frequent storm floods. This co-existence with the marshy landscape required some infrastructural adaptations, however. While some communities benefited from natural elevations in the landscape to build their villages, most had to adjust by creating elevations in the landscape – these mounds known as *Wurten* in Germany or *terpen* and *wierden* in the Low Countries. These mounds could be up to 6 meters above sea level and several hundred meters wide – sometimes encompassing the whole settlement. In this way, houses and warehouses were protected from the recurrent inundations, while the meadows, salt marshes, and creeks remained influenced by the tides.[17]

Inundated landscapes were not 'unaltered' pieces of wilderness either. To make sure the optimal amount of water entered the meadows and salt pans and the water was able to drain at the ideal time, sophisticated water management systems were developed. These systems should not be confused with embankments and drainage works: they managed the

---

[15] Baldwin, *Contagion and the State*; Briggs, 'Cholera and Society'; Cohn, 'Cholera Revolts.'
[16] Tarr, *The Search for the Ultimate Sink*, 182–184.
[17] Soens, Tys & Thoen, 'Landscape Transformation'; Van Dam, 'Denken over natuurrampen.'

Figure 4.1 George Pinwell, 'Death's Dispensary,' appeared in *Fun Magazine* (London, 1866). Pinwell's cartoon shows poor people surrounding a water pump that is being controlled by a skeleton or a figure of death, referring to John Snow's recent discovery about the cause of cholera.

flow and prevented an excess of water in bad weather. In the English Fenlands, for example, lodes (derived from the old English *gelād*, or to lead) and catchwater drains were installed from at least the early tenth century, but probably already by the seventh century. This elaborate system of pipes, canals, and ditches required extensive technical knowledge and the investment of ample time, labor, and capital. This gravity-controlled system regulated inundations and optimized hay production in the meadows.[18] In most regions around the North Sea, water

---

[18] Oosthuizen, 'Water Management'; Oosthuizen & Willmoth, *Drowned and Drained*.

management systems which fostered co-existence with floods were gradually replaced by other types of infrastructure which aimed to exclude flooding altogether and to introduce dryland agriculture in wetland environments. Only in some regions, such as the English Fenlands, could the amphibious system persist well into the seventeenth century, although increasingly marginalized by state-sponsored 'improvement' projects.[19] As these two technological systems served clearly different purposes – accommodating floods versus excluding them – they also produced profoundly different landscapes of risk and disaster.

4.2.2 Economic Pressures and Crises

The view that the pre-existing level of economic development determines the impact of hazards is perhaps most evident in the discourse about disaster vulnerability comparing the 'developing' world with the 'developed' world. From the nineteenth century onwards, but especially after 1945, disasters increasingly came to be seen as characteristic of a global South characterized by poverty, illiteracy, and backwardness. The West, by contrast, was believed to be largely protected from disasters by its high levels of economic and technological development.[20] In some regions this belief was supported by the relative infrequency of severe natural disasters for a longer period of time: Switzerland, for instance, experienced such a 'disaster gap' in the last quarter of the nineteenth century and the first three quarters of the twentieth century.[21]

There is a certain logic to the idea that economic preconditions, and more particularly poverty and 'underdevelopment,' affect hazard prevention, mitigation, and recovery. In affluent countries, higher living standards usually imply better-quality housing, a wider range of options for evacuation and shelter, and insurance to cover the costs of rebuilding after the disaster. Widespread poverty makes all of this much more difficult. In the 1970s, critical geographers like Phil O'Keefe and Ben Wisner framed disasters in the Third World, such as the 1976 Guatemalan Earthquake as 'classquakes' produced by underdevelopment and marginalization.[22] Poverty has also been singled out as the prime driver causing producers to abandon traditional restraints in the exploitation of fragile ecosystems. On the other hand, other scholars warn against the simple association of poverty with overexploitation of resources and exposure to natural hazards, since poor people may develop their own

---

[19] Robson, *Improvement and Environmental Conflict*.
[20] Bankoff, 'Rendering the World Unsafe.'   [21] Pfister, 'Die "Katastrophenlücke."'
[22] O'Keefe, Westgate & Wisner, 'Taking the Naturalness.'

coping strategies to minimize risks or mitigate the impact of natural hazards.[23] As well as living standards, public resources also matter: economically advanced countries are better positioned to generate, through taxation, considerable sums to invest in infrastructural works, warning systems, or other high-tech solutions aimed at preventing hazards or mitigating their impact. They may, however, not always be willing to do so. A variety of motives – ideological, political, social, economic – can prevent them from prioritizing hazard protection over other goals. Alternatively, they may be roused into raising expenditure only when it is too late, as the state of the levees in New Orleans before Hurricane Katrina suggests.

Economic development is not the only way in which economic preconditions may affect the impact of shocks, as the characteristics and structure of the economy – at local, regional, or national level – should also be taken into account. We focus here on three interrelated aspects: intensification, diversification, and commercialization. Even before the Industrial Revolution, intensification can often be linked to an increased exposure to natural hazards. In the eighteenth century, Europe was repeatedly afflicted by cattle plague, for instance, and its diffusion was directly connected to the intensification of the long-distance cattle trade in the preceding centuries. Every year, thousands of animals made the journey from their breeding places in Northern and Eastern Europe to the urban consumption centers in Western Europe. The dense concentration of animals provided the conditions for epidemic diseases, the outbreak of which was fostered by the vulnerability of the animals, which were often on the brink of starvation after the long journey West.[24]

Diversification can be seen as a risk-reducing strategy of medieval peasants. By growing multiple crops, cultivating many small plots scattered over a large area, and combining farming with non-agrarian activities such as gathering nuts and berries or cottage industries, the risks of adverse weather and subsequent harvest failure were reduced.[25] Research on famines in later periods suggests that regions with diversified economies fared better. During the major famine of the early 1590s in Northern Italy, the drop in the number of births in mountainous regions was not nearly as severe as in the lowlands, as mountain populations were able to supplement a grain-based diet with locally sourced dairy products, fruit and vegetables, and chestnuts collected in the woods,[26] while the

---

[23] Martinez-Alier, *The Environmentalism of the Poor*.
[24] Appuhn, 'Ecologies of Beef'; Brantz, 'Risky Business.'
[25] McCloskey, 'The Prudent Peasant' ; Pretty, 'Sustainable Agriculture.' See also Section 2.3.6.
[26] Alfani, 'The Famine of the 1590s.'

extreme dependency of the Irish population on potatoes was one of the factors – although not the only one – contributing to the dramatic impact of the potato blight in Ireland in the 1840s.[27]

The relationship between commercialization and vulnerability to hazards is more complex. On the one hand, in commercialized economies, substantial investments in hazard protection can be facilitated by the existence of markets for capital, labor, and commodities, if this serves the interests of the investors. This was the situation in Holland in the seventeenth and eighteenth centuries, when entrepreneurs investing in land reclamations were prepared to fund dikes and drainage systems in the expectation of substantial long-term returns.[28] Entrepreneurs, rural and urban alike, were not always keen on making large-scale investments in protection and mitigation measures, however. It depended on the timescale and the economic interests of the entrepreneurs. In Holland farmers were owners of the land and aimed for long-term profits, and as a result, commercialization was not accompanied by environmental pressures.

An unchecked quest for short-term profits, however, is more likely to lead to overexploitation of natural resources or to the neglect of protective elements. Indeed, commercialization can at times also be linked to accumulation practices and the constantly increasing production and consumption of renewable and non-renewable resources. A growing group of scholars has portrayed these economic strategies as clashing with environmental boundaries in a limited world. According to Moore, the spread of capitalism and commercialization created the preconditions for environmental disasters to occur throughout the world. Indeed, scholars from a wide range of disciplines are calling for a moral economy, degrowth, and a return towards commons and subsistence in order to prevent detrimental pressures.[29] Some historical sand drifts have been considered examples of the detrimental effects of commercialization, and the American Dust Bowl of the 1930s is often taken as a case in point.[30] Nevertheless, it remains unclear whether it was caused by a lack of environmental knowledge of the region or by the commercialization of grain production.

A related issue concerns the integration of the economy in international or interregional market networks. Open economies, at least in theory, have easier access to commodities and services in other regions, which may help them to cope with hazards and facilitate recovery, as has been

---

[27] Ó Gráda, 'Ireland's Great Famine.'
[28] Van Cruyningen, 'From Disaster to Sustainability.'
[29] Schneider, Kallis & Martinez-Alier, 'Crisis or Opportunity'; Raworth, *Doughnut Economics*; Moore, 'Environmental Crises and the Metabolic Rift.'
[30] Worster, *Dust Bowl*, 80–97. See also Section 1.2.

a long-standing argument for the relatively early escape from famine in England and the Northern Netherlands.[31] But integration in a network of markets may also result in the removal of scarce resources to other places. Markets, after all, respond not to needs but to purchasing power. The Irish Famine of the middle of the nineteenth century is a case in point again. Food was imported to Ireland, in the form of sizable quantities of maize that arrived directly from America in the spring of 1847. Although this was not enough, it helped to save lives. But in the winter before the maize arrived, grain exports from Ireland to England had taken place in much the same way as they had done before the crisis, "presumably because the poor in Ireland lacked the purchasing power to buy the wheat and oats that were shipped out."[32]

While this book does not focus on wars as hazards or disasters in themselves, we should also note that conflict could affect and exacerbate the impact of other environmental hazards. Only in very recent times have we seen how distrust of health workers, clinicians, governments officials, and drivers during Ebola outbreaks in Western Africa – sometimes leading to violence – was connected to pre-existing levels of suspicion and distrust linked to ongoing civil war conditions.[33] The difficulties of dealing with hazards during wartime are also seen going back into the pre-industrial past: in medieval Flanders, for example, the Flemish–French war of 1314–15 formed one of the main preconditions for the harvest failures of 1315–17 – caused by excessive rainfall – to turn into a full-blown famine.[34] In order to finance the war and feed the troops, the Count of Flanders implemented a confiscation policy, redirecting the goods and grain of rebels and political adversaries to the national treasury and army. Most of these confiscations of grain were derived from the front zones in Flanders: the towns of Ypres and Cassel. Other regions escaped relatively lightly and could contribute in cash rather than grain, which was an advantage, given the harvest failures. To make things worse, the provisioning of the army interfered with local grain markets. In areas where troops were stationed, extra grain was purchased, which meant ever lower grain stocks available for the local population. The usual famine mitigation measures of grain imports were prevented as well, since the County of Flanders derived most of its imported grain from the North of France, a trade route now blocked because of impending

---

[31] But see Curtis & Dijkman, 'The Escape from Famine.'
[32] Ó Gráda, 'Ireland's Great Famine,' 53.
[33] Blair, Morse & Tsai, 'Public Health and Public Trust'; Wilkinson & Fairhead, 'Comparison of Social Resistance.'
[34] Geens, 'The Great Famine.'

hostilities. Because of these burdens of war, Ypres and Cassel were the worst-hit regions within the County of Flanders during the Great Famine.

4.3 Coordination Systems and Institutional Preconditions

Institutions are highly relevant to the issue of vulnerability and resilience. Broadly defined as formal and informal rules and the associated organizations and networks, institutions can be specifically designed and implemented to cope with hazards: relief organizations, emergency legislation, forms of insurance, or rescue systems, to name but a few.[35] The functioning of these specific hazard-oriented institutions is central to the study of historical disasters. However, it has become increasingly clear that the ability of communities and societies to cope with hazards depends not only on these specific institutions, but also on the overall 'ordinary' institutional infrastructure. Among them are the institutions which organize the exchange, allocation, and use of resources more generally; for instance, the arrangement of property rights or the institutions structuring market exchange. Sometimes indirectly, but often also directly, they affect the capacity of societies in preventing hazards turning into disasters, or their capacity to recover quickly.[36] Institutions usually do not function in isolation. They are embedded within one of the larger coordination systems that regulate the allocation of resources in any society: the family, the state, the market, and various forms of collective action. The preconditions shaped by these coordination systems are the main focus of this section.

The extent to which institutions affect the capacity of societies to cope with shocks is debated in the literature, although not always in a systematic way. Viewpoints in the debate are often loosely related to the varying views on the formation of institutions more generally. Sometimes institutions are perceived as the result of rational choices made by utility-maximizing individuals. In this view, competition ensures an optimal outcome: the best and most efficient institutions survive.[37] This position, however, is hardly tenable in the light of the persistence of a multitude of institutions that increase vulnerability instead of reducing it. Indeed, institutions often have oppositional effects. While an institution might, for instance, help to push up profitability or clarify and secure property rights, it might also at the same time have detrimental effects on sustainability – thereby increasing vulnerability. Furthermore, it is

---

[35] See for these institutions and organizations the extensive discussion in the next chapter on prevention and responses.
[36] Bankoff, *Cultures of Disaster*, 11–13. [37] North, *Institutions*, 17–20.

precisely the profitability for some that may obstruct attempts of others to adapt the institutions in question in order to reduce vulnerabilities or enhance resilience. This problem, connected to the multifaceted effects of institutions, thus points to a more fundamental issue: institutions are often the result of social bargaining or even conflict. They are intimately linked up with the leverage and positions various social actors have and may therefore be formed and dictated by the interests and preferences of certain individuals and social groups.[38]

The embeddedness of institutions in social, political, and economic structures also explains why many institutions cannot easily be changed: once they are in place, they tend to be reinforced by the groups that profit from them. In cases like this change takes place only under great pressure. Sometimes, disasters themselves can create such pressures. A study of recent floods in the Netherlands and in Poland shows that hazards and disasters can create a perfect window of opportunity for institutional change. As 'focusing events' they draw attention to risks and may emphasize the urgency of action – in turn leading to institutional change.[39] Mostly, however, institutions are the result of a path-dependent process and not easily changed; that is, they have an entire logic of their own. The capacity of societies to cope with hazards through their institutions, therefore, cannot be taken as a given, as they will not automatically be geared towards coping but will instead be geared towards the interests of certain interest groups, or can persist even if they weaken a society's coping capacity. Institutions devised to tackle one challenge may also have side-effects (positive or negative) in other domains, which are often largely unforeseen. Some of these themes are now explored in the sections that follow.

4.3.1 Coordination Systems: The Family, the Market, and the State

All societies require some form of coordination to organize the exchange and allocation of resources. Four main systems can be identified: the family or household, collectives such as local communities and associations, the market, and the state. These systems do not primarily develop with the explicit intention of coping with hazards, shocks, and disasters, but through their sets of political, social, and economic institutions they do affect a society's capacity to mitigate the impact, recover, or prevent recurrences. With the possible exception of small groups of hunter–gatherers, all societies rely on more than one coordination system at the same time. The exact combination, however, varies widely, both between

---

[38] Nee & Ingram, 'Embeddedness and Beyond'; Ogilvie, '"Whatever Is, Is Right"?'
[39] Kaufmann et al., 'Shock Events and Flood Risk Management,' 51.

societies and over time. All these coordination systems have changed fundamentally through time and with varying trends and evolutions over the globe. Processes such as state formation and state failure, the development and demise of formalized collectives, and the shift between the core household and extended family systems have significant repercussions for vulnerability and resilience towards hazards.

When struck by a calamity, many people turn to family members first for basic necessities such as shelter, food, and emotional support. Families can also be instrumental for recovery afterwards. Families are great providers of 'bonding' social capital: horizontal ties within homogeneous groups that generate solidarity and reciprocity.[40] Although the European marriage pattern, characterized by a predominance of the neolocal nuclear household, consensual late marriage, and monogamy, has been viewed positively for lowering child mortality rates, raising education, and stimulating economic development,[41] its specific impact on vulnerability towards hazards and shocks is unclear in the absence of convincing micro-level research.

On the one hand, as demonstrated by research in the aftermath of hurricane Katrina, household size affects evacuation behavior. In New Orleans, high costs and practical problems related to the needs of elderly family members complicated evacuation decisions, and consequently large extended households were more likely to stay behind than small nuclear ones.[42] Extended families, on the other hand, offer other advantages. In the summer of 1994, torrential rains during a typhoon caused a terrible flood in the Beijiang basin in the southern province of Guangdong, China. Although there were no casualties, the flood caused great damage. Summer crops were almost entirely destroyed, and in the city of Qingyuan and the surrounding villages more than 80 percent of the houses collapsed.[43] Even though aid, in the form of bricks and funds, was provided by donations from Hong Kong, the state was not able to mitigate all effects and propagated self-reliance. Kinship, social networks, and communal ties therefore became crucial. During the rebuilding period the Chinese victims of this great flood relied on these ties not just for food and shelter, but also for loans in order to buy the seed and equipment they needed to resume farming – with extended families acting as a buffer.[44]

---

[40] Putnam, *Bowling Alone*, 22–24.
[41] De Moor & Van Zanden, 'Girl Power'; Foreman-Peck, 'The Western-European Marriage Pattern'; for a dissenting view: Dennison & Ogilvie, 'Does the European Marriage Pattern Explain Economic Growth?'
[42] Tierney, 'Social Inequality,' 116–117.
[43] Wong & Zhao, 'Living with Floods,' 190, 193.
[44] Wong & Zhao, 'Living with Floods,' 198.

Marriage patterns could have very different effects on vulnerability levels depending on the strength of marital ties, as is shown by Megan Vaughan's study of a famine in lowland Nyasaland in 1949. Vaughan stresses that entitlements to food of households are only part of the story: it is also important to look at allocation and distribution of food within the household too. In lowland Nyasaland it was customary for women to marry men from the highlands in the North and West, who would then move to the lowland to live with their wife and her family. In 1949 this custom brought advantages: many husbands made the arduous journey to their relatives in the highland region and were able to bring back food supplies to their wives and children. Others, however, stayed with their relatives until the famine was over, and some even found themselves a new wife in their region of origin, leaving their first spouse destitute.[45] This marriage pattern could thus have very different effects on vulnerability levels, being highly dependent on the strength of the marital ties and the willingness of marital partners to extend relief.

In almost all societies, past and present, yet another coordination system is at work: the market. Commodity markets are usually the first to arise; well-developed markets for land, labor, and capital are, from an historical perspective, not as common.[46] Through their role in allocating resources, the existence and the organization of markets affects the coping capacity of societies during crises. Notably, Amartya Sen's influential analysis of the Bengal famine of 1943 is largely based on the combined effects of labor and commodity markets. Sen argued that the famine was not a question of insufficient food, but rather of specific groups experiencing insufficient access to food. The reasons for this 'entitlement decline,' he claimed, originated partly from 'market imperfections' related to wartime inflation, hoarding, and speculation, but also from the fact that the wages of certain occupational groups stagnated, or even fell, while the prices of basic foodstuffs rose.[47]

Exactly how markets affect resilience and vulnerability depends, again, on political, social, and economic circumstances. In fifteenth- and sixteenth-century coastal Flanders, growing commercialization in combination with changing property relations gave rise to reduced flood protection. With the increase of urban landownership and the reduction of peasant smallholding, investments in water management fell. Peasants prioritized safety and continuity of their holdings, but absentee urban landlords adapted their strategies to expectations of profitability: they were not eager to step up investments in the maintenance and

---

[45] Vaughan, 'Famine Analysis,' 186–189. [46] Van Bavel, *The Invisible Hand?*
[47] Sen, *Poverty and Famines*, 63–70.

4.3 Coordination Systems

improvement of dikes, drains, and sluices, as risks were high and returns, in the shape of lease prices, modest.[48] In contrast, in seventeenth- and eighteenth-century Holland and Zeeland, urban entrepreneurs were quite willing to invest in equally risky large-scale reclamation and re-embankment projects, and thus contributed to the improvement of flood protection. Investments were stimulated by expectations of substantial profits, but also by the fact that the political power of urban entrepreneurs allowed them to negotiate tax exemptions and thus reduce costs.[49]

The last coordination system to be discussed here is often referred to as the 'state': a convenient shorthand, but not a very accurate one since various local and regional authorities also reside under this heading. How governmental bodies prepare for and respond to hazards is important; their role in this respect will be discussed in more detail in Chapter 5. Just like family systems or markets, governmental systems may affect the capacity of societies to deal with shocks. In modern societies, individuals have high expectations of government when it comes to disaster management. However, efforts by governments to prepare for shocks or mitigate their impact are not new. Some governments went to considerable trouble to set up warning systems. The thirteenth-century caliphs of Baghdad, for example, maintained a system to warn the city of an impending flood by having the water level of the Tigris measured at a location hundreds of kilometers upstream.[50] Another kind of preparation concerns emergency relief, the organization of which tends to reflect prevailing political and social relations. In Europe, with its tradition of urban self-rule, famine relief was first and foremost a responsibility of town governments. Between the fourteenth and the sixteenth centuries, urban authorities in many parts of the continent, driven at least in part by fear of civil disturbance, established public grain stocks to relieve distress in cities.[51] Here, however, the focus is on the conditions created by pre-existing governmental systems not specifically aimed at hazard and disaster management.

The belief that regime type impacts coping capacity is at the heart of Amartya Sen's claim that democracy provides a safeguard against famines. According to Sen, a free press and political accountability will force authorities to take decisive and timely action to prevent the worst.[52] Other researchers have focused not so much on democracy per se as on specific

---

[48] Soens, 'Floods and Money.' [49] Van Cruyningen, 'From Disaster to Sustainability.'
[50] Weintritt, 'The Floods of Baghdad,' 168.
[51] For example in Italy: Alfani, *Calamities and the Economy*, 70–78; in the North Sea region: Dijkman, 'Coping with Scarcity.'
[52] Sen, *Development as Freedom*, 178–184.

institutions and practices of good governance found more frequently – but not exclusively – in democratic societies, such as high government capacity, controls on corruption,[53] effective decentralized government,[54] checks to prevent domination of government by a single interest group, a clear division of responsibilities between governmental bodies,[55] and a combination of political, legal, economic, and social mechanisms and pressures to ensure that mobilization against famine actually takes place.[56]

How vulnerability and resilience can be affected by a failure to fulfill these conditions is demonstrated by the famines that occurred in Ethiopia and Sudan in the 1980s. These famines may have been triggered by drought, but they were exacerbated by the lack of political voice among the poorest and by the suppression of the press. Initiatives to address the emerging crises were not sufficiently supported by the government or were used for political goals.[57] In Ethiopia, the dictatorial socialist Dergue regime played a particularly damaging role by imposing heavy grain exactions on the countryside in order to feed the army and the urban population, and especially by the campaign to counter insurgency in the Tigray region. This campaign included the destruction of crops and rural markets, the imposition of restrictions on trade, forced resettlement of people, and the denial of relief to the region.[58]

Institutional preconditions can also be influenced by other characteristics of the state than regime type, such as the ideological underpinnings of state power. In early-modern Europe, and increasingly so during the seventeenth and eighteenth centuries, changing attitudes to the control of nature and the role of knowledge and expertise gave rise to a much more proactive role of the state. A political culture of stewardship induced states to focus on protecting their subjects from the vagaries of nature, as for example in the construction of dikes and retaining walls on the Canal du Midi in seventeenth-century France. Using techniques derived from military engineering, dikes and retaining walls shielded the land from flooding, thus emphasizing the ability of the king to tame the 'wild weather' and control the land. This, in turn, reinforced the legitimacy and authority of state power: disaster management and state building went hand in hand.[59] The eighteenth century shows us the first examples of state intervention based on the compilation of 'scientific data,' as in eradication policy in several European states to deal with rinderpest, or the reconstruction of Lisbon after the earthquake of 1755.[60] Underneath

---

[53] Burchi, 'Democracy, Institutions and Famines.'
[54] Banik, 'Is Democracy the Answer?' [55] Rubin, 'The Merits of Democracy.'
[56] De Waal, *Famine Crimes*, 11. [57] Von Braun, *A Policy Agenda*, 8.
[58] De Waal, *Famine Crimes*, 110, 116–120. [59] Mukerji, 'Stewardship Politics,' 129.
[60] Van Roosbroeck & Sundberg, 'Culling the Herds?'; Araújo, 'The Lisbon Earthquake.'

4.3 Coordination Systems

this discourse of the 'common good,' severe differences of opinion could be present, as in the case of the drainage of the Fens in England, where the state's desire for control over nature clashed with the local population's survival mechanisms.[61] Whether heightened state intervention is seen as a success story or as more efficient is therefore clearly in the eye of the beholder.

In the modern era centralization of political power, expanding state resources, and the rise of the welfare state have contributed to a vigorous increase of state intervention in disaster management. In the late nineteenth century and early twentieth century, when new 'professional' disaster relief organizations, such as the Red Cross, started to operate alongside private philanthropic networks, the state was still largely absent. The Cold War era marked the rise of national civil defense in the Western world, as the threat of air raids and particularly of nuclear attack grew. The legal framework this created played its part in the emergence of national disaster management and relief schemes.[62] International disaster relief schemes emerged from the 1970s onwards, as the UN was increasingly criticized for its lack of organized response when disasters struck in developing countries. The United Nations Disaster Relief Organization (UNDRO) was established in 1971/72, and disaster relief and management became a prerequisite at the national level for member states as well.[63] For the same reasons, however, the chances of governments becoming part of the problem instead of its solution have also increased. International disaster relief has also received fierce criticism, as relief organizations often end up empowering authoritarian regimes and disempowering victims, linking up with Sen's focus on the importance of democracy and political contract.

4.3.2 Institutions for Collective Action and the Commons

The family, the market, and the state are the three most familiar coordination systems, but in many societies their role was complemented by other forms of cooperation and collective action – sometimes assuming a prominent role in relation to hazards and shocks. Family resources to deal with hazards, for example, were often restricted by their size, scope, and capacity to last, and thus relief strategies were often supplemented by coping mechanisms that relied on collective action: various forms of cooperation between individuals and households. Here, another type of social capital comes into focus: bridging social capital links people from

---

[61] Ash, *The Draining of the Fens*. [62] Quarantelli, 'Disaster Planning.'
[63] Kent, 'Reflecting.'

different social backgrounds, giving them access to assets from a wider segment of society.[64]

Cooperation and coordination are not always institutionalized: informal networks, for instance between kin or neighbors, may also perform these functions. For some scholars, however, formalized collective action in particular is a vital component of risk reduction vis-à-vis hazards. This is argued for pre-industrial Northwestern Europe, where the nuclear household was the basic family unit from the Middle Ages onwards. In what has been labeled the 'Silent Revolution,' from the twelfth and thirteenth centuries onwards, collective action became formally institutionalized in organizations such as commons, guilds, fraternities, water boards, and beguinages, in order to reduce risks, share costs, and offer broader welfare and protection opportunities.[65] These institutions offered solidarity, where vulnerable community members received aid to secure their livelihood. Guilds, for instance, could protect small entrepreneurs from fluctuating markets and volatile prices.[66] Beguinages provided a secure life for single women, surrounded and supported both by blood-kin and by non-relatives.[67] Religious or civic poor relief organizations funded by charity offered support to those unable to provide for themselves: the poor, the sick or handicapped, the elderly – even if in comparison with modern welfare states the levels of assistance provided by these institutions were modest.[68]

Most institutions for collective action were not primarily established for the purpose of managing hazards or the ensuing disasters. Instead, they addressed structural issues, which indirectly impacted levels of vulnerability and resilience. During famines, for instance, poor relief organizations faced a growing demand for assistance: people who survived on the edges of subsistence in normal times were threatened when food prices peaked. In the sixteenth-century Southern Low Countries, food distribution by the 'poor tables,' organizations run by members of the local community, increased significantly during major food crises. Nevertheless, context mattered: distributions were more forthcoming in regions with a more equitable distribution of power and wealth, as a broader base of people donated to the system in normal years as a form of 'collective insurance.'[69] However, financial reserves of poor

---

[64] Putnam, *Bowling Alone*, 22.
[65] De Moor, 'The Silent Revolution'; de Moor, *The Dilemma of the Commoners*.
[66] For this kind of view on guilds see de Munck, *Guilds*; de Munck, 'Fiscalizing Solidarity'; Prak *et al.*, *Craft Guilds*. Although at the same time potentially benefiting members to the detriment of non-members: Ogilvie, *The European Guilds*.
[67] Overlaet, 'Replacing the Family?'; de Moor, 'Single, Safe, and Sorry?'
[68] Van Bavel & Rijpma, 'How Important Were Formalized Charity and State Spending?'
[69] Van Onacker & Masure, 'Unity in Diversity.' See also Section 4.3.1.

relief organizations were often limited, as was their ability to raise additional funds, especially during times of general hardship. Many of them were therefore unable to sustain significantly raised levels of expenditure over a long period.[70]

Notably, in several countries – most notably (rural) France and Spain – institutionalized relief during famines was almost entirely absent. In the French countryside, food security was often guaranteed by the communal practice of gleaning.[71] Outside of Europe, famine aid also often took shape in different ways: in eighteenth- and nineteenth-century India, for example, the common form was individual giving by well-to-do landowners or merchants in the form of food, alms, or work. This kind of charity was frowned upon by the British, who objected to its indiscriminate character: they feared it would pauperize people and make them dependent on aid permanently. This thinking was very much in tune with the developments in Great Britain itself, where the 1834 Poor Law, prohibiting outdoor relief in favor of the Victorian workhouses, was even more focused on labor control than earlier legislation.[72] Yet, with the rise of civil society in the late nineteenth century, indigenous Indian famine relief was partly transformed as voluntary local committees emerged to raise funds, start kitchens, or organize work projects.[73] Overall, however, the impact of formalized relief must not be overstated – even in Western Europe. While some forms of collective action were successful in lowering vulnerability levels, others had only a minor impact on the members of the collective.

Among the institutions for collective action, commons deserve special attention: they have been seen as an effective institutional framework for reducing vulnerability by limiting environmental degradation and offering buffers against, for instance, soil erosion and sand drifts.[74] In some contexts, arable agriculture was organized collectively. Open fields, a basic system of scattering of arable plots across a number of fields, offered a number of risk-limitation possibilities: preventing certain landholders from monopolizing the most fertile soils, spreading the risk of drainage problems and overall crop failure, and equitably distributing the distances needed to walk from settlement to fields. This kind of tactic is still seen around the globe today in peasant societies. Another form of

---

[70] Dijkman, 'Feeding the Hungry.' Similarly, crises such as epidemics also limited the dedicated guild funds to assist members during hardship: Van Leeuwen, *Mutual Insurance*, 17–82.
[71] Vardi, 'Construing the Harvest.' [72] King, 'Welfare Regimes,' 56.
[73] Brewis, 'Fill Full the Mouth of Famine,' 890, 897, 901.
[74] De Moor, 'Avoiding Tragedies'; de Moor, 'Participating'; Beltrán Tapia, 'Social and Environmental Filters'; van Zanden, 'The Paradox of the Marks.'

commons was the shared, but rationed and regulated, access to rights and obligations over non-arable resources such as pastures, forests, wastes, and marshes. Such rights have been seen as an important form of welfare or protection for pre-modern rural inhabitants, especially if they had little access to arable land, and could benefit from the right to graze animals, to pick herbs, fruit and fungi, to hunt and fish, and to extract fuel and building materials such as dung, timber, and peat.[75]

The idea that the commons could bring about reductions in societal vulnerabilities has not always been accepted. The most influential and widely cited critic was ecologist Garrett Hardin, whose story of the 'tragedy of the commons' argued for a number of dangers – in particular the problem of avaricious individuals acting in self-interest who would eventually deplete finite common resources or goods.[76] This would eventually lead to a 'tragedy of the commons,' in the form of soil degradation, a shift of the ecological stability domain, or a subsistence crisis because of a lack of natural resources. However, in more recent times, the commons as an effective tool for managing resources has been re-established by scholars pointing to the fact that the use of common resources was normally regulated and sanctioned.[77] In many cases formal restrictions applied, dictating how much of a resource could be used, and, more importantly, by whom. In that sense, commons were no 'free-for-all' doomed to ruin by greedy individuals, but instead were complex, multi-layered, adaptable, and often exclusionary.[78] The fact that evidence exists for societies throughout history passionately defending their collective rights to different resources points to a system that continued to offer many people a large number of real benefits in terms of reducing vulnerabilities.[79]

The disintegration of the commons – in various parts of Europe basically complete by the nineteenth century – is now often interpreted as a negative development, especially for the poor, who were more reliant on these resources. There is a parallel in the contemporary world, where the poor are disproportionately dependent on the commons in developing and underdeveloped countries.[80] Also, many scholars point to a rise in vulnerability levels more generally after the decline of forms of collective action, as has happened in current-day African wetlands, where the loss of traditional rights to fishing as a common property resource has eroded the

---

[75] See also Curtis, *Coping with Crisis*, 40–42. [76] Hardin, 'Tragedy of the Commons.'
[77] Ostrom, *Governing the Commons*; Casari, 'Emergence of Endogenous Legal Institutions,' 220; de Moor, 'Avoiding Tragedies'; Laborda Pemán & de Moor, 'A Tale of Two Commons,' 13.
[78] Congost, 'Property Rights,' 90. [79] Curtis, 'Did the Commons Make,' 650.
[80] Beck & Nesmith, 'Building on Poor People's Capacities,' 119; Jodha, *Life on the Edge*.

4.3 Coordination Systems

livelihoods of local communities.[81] The privatization of commons has even been labeled as (common) land grabbing and resource grabbing. Where communities in Morocco and Ghana were hitherto able to cope with periods of crisis thanks to access to communal land and collective grazing, the privatization of those communal resources and the abolition of collective action significantly impacted on their level of vulnerability to hazards and crises.[82]

The increasing realization now is that the potential for collective action in general and the commons in particular, both past and present, to reduce vulnerabilities stands somewhere between the completely negative views of Hardin or the 'Enlightenment reformers' and the wholly positive interpretations now being spun by those working directly within the 'commons' or 'collective action' sub-fields. For the pre-industrial period at least, it would be wrong to think that the commons were equitably divided and the poor always had sufficient access to welfare and protection components.[83] In fact, historically speaking, this was more the exception than the rule. Indeed, mirroring broader inequalities across societies in general, especially as we move into the early-modern period, we find commons as part of restrictive access regimes, with highly stratified access and fear of encroachment by outsiders.[84]

Indeed, within the commons, access to collective rights could be attached to privileged farms or families, acceptance or integration into the community, length of residence or lineage, or payment of license fees, or related to ownership of private land and livestock.[85] That is to say, accumulation of land could also mean a decreasing number of residents with actual access to the commons or a say in how they functioned. Increasing levels of inequality could thus be detrimental to the good performance of commons, manifesting themselves in ecological problems such as the seventeenth-century sand drift that destroyed the village of Santon Downham and clogged the nearby river in the English Brecklands.[86] Parallels can be seen with highly stratified and exclusionary access to other collective institutions. Decision-making in medieval water boards, for instance, was traditionally restricted to landowners, and in the Flemish coastal area, for example, this category originally included a large segment of the region's population. However, from the thirteenth century onwards, when smallholding made way for tenant farming, the water boards of the Flemish coastal area were increasingly dominated by

---

[81] Haller, 'Understanding Institutions.'
[82] Ryser, 'Moroccan Regeneration'; Gerber, 'New Commons and Resilience.'
[83] Curtis, 'Did the Commons Make.'   [84] De Keyzer, 'The Impact,' 521.
[85] On all these different methods: Curtis, 'Did the Commons Make.'
[86] De Keyzer & Bateman, 'Late Holocene Landscape Instability.'

absentee landlords, while local peasants were hardly represented at all. The result was a decline in investments in water management, ultimately giving rise to increased vulnerability to flooding.[87] In general, therefore, the historical evidence suggests that there is a link between collective action and a reduction in vulnerability – although still more important for the outcome was the particular socio-political context in which the collective institution operated.

4.4 Social Pressures: Poverty, Inequality, and Social Distress

Poverty and inequality in the distribution of wealth, resources, and incomes are often cited as significant factors influencing vulnerability – which is important, given that these two factors have shown strong variation between societies and across time. Do high levels of poverty and economic inequality affect societies' capabilities to anticipate shocks and hazards, mitigate the effects of disasters, and adapt to them? Indeed, poorer individuals and groups are said to be most severely affected by hazards and disasters, and a number of aspects of 'being poor' are said to contribute to this. The poor tend to live in inherently hazard-prone locations, they lack the capital to invest in preventive measures or build up resource buffers for anything unexpected, they have more restricted access to helpful social networks, and frequently are disenfranchised from the political process that can help steer policies more conducive to their protection.[88] They also tend to have lower standards of health through poorer diets and access to healthcare. Drought is the hazard said to be most clearly connected to cases of extreme poverty.[89] Income, wealth, and access to material resources are also significant factors in explaining why certain communities are hit hardest by hurricanes.[90] In New Orleans after Hurricane Katrina in 2005, the poorest groups, disproportionately coming from African-American communities, inhabited the lowest and most flood-prone parts of the city, while wealthier communities lived on the land near the river front that was 3 meters higher.[91] Indeed, in the United States, it is often the case that poorer ethnic minorities are disproportionately located in inferior housing physically segregated into low-value neighborhoods. It is this segregation that creates so-called 'communities of fate,' whereby residents share the same fate regarding quality of life and opportunities, but also regarding exposure to certain types of hazards.[92]

[87] Soens, *De Spade in de Dijk?*, esp. Chapters 2 and 3.
[88] Sen, *Poverty and Famines*; Wisner et al., *At Risk*.
[89] Shepherd et al., *The Geography of Poverty*.   [90] Reed, 'The Real Divide,' 31.
[91] Colton, 'Basin Street Blues,' 237.   [92] Logan & Molotch, *Urban Fortunes*, 19–20.

4.4 Social Pressures

The issue of the impact of inequality on vulnerability is more complex, however, and discussed less frequently. One of the major issues is the lack of explicit distinction between the effects of unequal distribution of wealth and resources, on the one hand, and overall poverty, on the other. For example, Ted Steinberg in his book *Acts of God* assumes that more equitably arranged societies can reduce the destructive effects of hazards,[93] and yet, within his analysis, it is difficult to distinguish between the negative effects of poverty and those of unequal distribution – especially when both exist simultaneously. Recently, statistical analyses have established a positive correlation between income inequality and increased susceptibility to disastrous outcomes from natural hazards in present-day countries.[94] Controlling for the number of natural disasters and national wealth, countries with less income inequality (as well as more democratic nations) suffer fewer deaths from disasters. This effect has been observed for a set of fifty-seven countries analyzed for the period 1980–2002, with the effect of income inequality found to be very large.[95] Empirical research at meso- or micro-levels is rare, however, and studies at the household level on the impact of wealth on coping strategies have provided inconclusive results.[96] While poverty may hamper adaptive capacities,[97] inequalities often remain undiscussed. Moreover, we lack insight into the exact aspects of inequality that make a society less able to cope or adapt. One of the reasons for this may be that different kinds of inequalities can impact differently upon vulnerability – the effects of income inequality may not be the same as the influence of differences in the distribution of resources, or of the differential access to voting rights and networks.

Long-term historical research may help us identify some of the mechanisms that are at play when linking vulnerability and inequality: history can function as a laboratory, where the effect of different types of inequality interacting with the same type of hazard can be tested in a long-term perspective.[98] Some evidence from history suggests that while the links between poverty and vulnerability are often direct – by dictating habitation locations, resource buffers, social networks, or exposure to unpredictable markets – the links between inequality and vulnerability are often indirect and therefore more complex. During a series of floods in the pre-

---

[93] Steinberg, *Acts of God*.   [94] Hillier & Castillo, 'No Accident,' 16.
[95] Kahn, 'The Death Toll.'
[96] Hoddinott, 'Shocks and Their Consequences'; Béné *et al.*, 'Is Resilience Socially Constructed?'
[97] Carpenter & Brock, 'Adaptive Capacity'; Carter *et al.*, 'Poverty Traps.'
[98] Van Bavel & Curtis, 'Better Understanding Disasters'; Curtis, van Bavel & Soens, 'History and the Social Sciences.'

modern Low Countries, for example, it has been shown that economic inequality – defined in this case as unequal ownership of wealth or property and use of resources – especially affected societal responses to hazards and shocks by impacting upon the development and use of institutions – in this case water boards.[99] Vulnerability, as measured by increasing prevalence and severity of flood outcomes, tended to occur when water management institutions failed to adapt to a context of redistribution of economic resources. Many medieval water boards functioned well under a system that attributed the construction and maintenance of flood defenses to a large group of smallholders: each was individually responsible for maintaining a part of the dike or drainage system, according to the size of the holding. However, when at the end of the Middle Ages or in the early-modern period – the timing varied between regions – landholding was gradually consolidated in the hands of elites, fewer people had an incentive to contribute their labor or money.[100] The quality of flood defense deteriorated significantly.

Very similar kinds of movements towards greater economic inequality, together with an insufficient regulation of actions of the elite, have been associated with dysfunction in water management structures in contexts as diverse as Mamluk Egypt in the fifteenth century and the British Punjab of the late nineteenth century.[101] Comparable effects can also be demonstrated for other types of disasters, for instance through the impact of drought in pre- and early-colonial Southern Africa.[102] Despite increasing inequality, seventeenth-century expansion of Portuguese landholding in the lower Zambezi valley brought with it an increased diversity of cultivated crops, including winter wheat, which reduced the sensitivity of the agricultural system to drought during the main summer rainy season. This, together with access to grain imports, centralized grain storage, and localized decision-making, enabled increasingly effective responses to short-term drought events. In the eighteenth century, however, continued growth in absentee landownership, together with the concurrent growth of the slave trade, led increasing numbers of landholders to seek short-term gains by selling peasant farmers to coastal slave traders, but at the expense of the core agricultural functioning of the estates. Initial reductions in the sensitivity of African peasant farmers to short-term low-magnitude drought events through crop diversification therefore masked gradual but fundamental new vulnerabilities, which

---

[99] Van Bavel, Curtis & Soens, 'Economic Inequality.'
[100] This has also been shown to be the case in the Po Valley of Northern Italy in the sixteenth century: Curtis & Campopiano, 'Medieval Land Reclamation.'
[101] Ali, 'Malign Growth?,' 124; Borsch, 'Environment and Population.'
[102] Hannaford & Nash, 'Climate, History, Society.'

4.4 Social Pressures

were exposed when a severe multi-year drought in the period 1824–30 led to the near breakdown of the entire social and agricultural system.

However, inequality was not inevitably a barrier to reducing vulnerability. To return to flood protection in the pre-modern Low Countries: in some regions a new balance was reached at a later point in time, on the basis of the investment of large amounts of capital by wealthier absentee elites in a more unequal setting. This did work well, but it materialized only once the institutional system had shifted definitively to a fully commercialized and monetized system based on contracted wage labor.[103] Accordingly, both equal and unequal societies could produce reduced levels of vulnerability. The same complexity has been demonstrated by research into inequality in the pre-modern Brecklands of Southeast England – a fragile ecosystem with inherent pre-existing vulnerabilities to erosion and sand drifts. Here, polarization in the distribution of wealth did not necessarily lead to a higher level of vulnerability. Only when this was accompanied by political inequalities, with elites such as landlords and wealthy tenant farmers monopolizing local decision-making, did the Breckland communities become more exposed to sand drifts.[104]

What could be said, then, is that a system of inequality was sometimes compatible with low vulnerability, if reciprocal agreements were established between elites and those with fewer resources that enhanced welfare and protection.[105] The extent to which this happened, however, depended on the precise incentives of those elites, and whether their power and prosperity were intrinsically related to and dependent on the welfare of the poor and the reproduction of the institutions that protected the interests of broad groups – in a sense creating a type of 'collective wealth.'[106] This was a situation characteristic of many pre-modern societies with patron–client relationships at the core of social stability and economic well-being,[107] and, more specifically, a situation that characterized many feudal societies where the power of the lord was not simply vested in ownership of large amounts of land and capital, but was also dependent on (and limited by) the efforts of the peasantry to work this land and pay rents and taxes.[108] Indeed, a main source of vulnerability in England during early-fourteenth-century subsistence crises and the famine of 1315–17 was perhaps not the inequalities between lord and peasants, but rather those in the ranks of the peasantry themselves. Wealthier tenants established more secure positions for themselves by consolidating

---

[103] Van Bavel, Curtis & Soens, 'Economic Inequality.'
[104] De Keyzer, 'The Impact of Inequality.'   [105] Levi, 'Aequitas vs Fairness.'
[106] Di Tullio, 'Cooperating in Time of Crisis.'
[107] In late Qing and early Republican rural North China, for example: Duara, 'Ten Elites.'
[108] Curtis, *Coping with Crisis*, 57.

the property of those weaker tenants who had to sell land out of desperate necessity.[109] If it suited their purposes, elites could cooperate with lower socio-economic groups for mutual benefit, investing in works and coercing others to build water management structures,[110] construct defensive and protective infrastructures, and perform obligatory public works strengthening agriculture.

4.5   Cultural Preconditions

Within the field of disaster studies, increasing attention is being paid to cultural factors. Worldviews, values, norms, attitudes, and customs shape the capacity of communities to cope with shocks, while disasters in turn may shape culture. Although this section focuses on pre-existing cultural factors, in regions characterized by high-frequency hazardous events, it could be said that culture and risk co-evolve over time.

How cultural preconditions may affect coping capacity is demonstrated most clearly, perhaps, by the multi-faceted role of religion, defined as a coherent system of beliefs, values, practices, and organizations. Even in modern societies religion is often invoked when a disaster occurs. Ted Steinberg has, for instance, pointed out that in the United States disasters are frequently presented as 'acts of God.' According to him, this exonerates political leaders from the responsibility to prevent those calamities, and allows them to refrain from addressing the social and economic inequalities that explain the high vulnerability of specific groups in American society.[111] However, religious beliefs can also shape the way in which individuals and communities perceive hazards, shocks, and disasters, for instance as divine punishment for wrongdoings, or as a test of faith. Such perceptions subsequently affect coping strategies. If a sudden shock is perceived as the conscious action of a displeased deity, it makes good sense to aim for appeasement by reaching out to the supernatural via a religious ritual. While from a secular perspective it is tempting to think of such a course of action as ineffective, religious rituals can bring comfort, hope, and a sense of belonging. Religion may also affect vulnerability of groups or individuals negatively, in more direct ways: by prescribing certain practices and forbidding others. Dietary and hygiene regulations may, for instance, affect susceptibility in case of epidemics, positively or negatively. Moreover, through its institutions, religion can contribute significantly to the potential for collective action:

---

[109] Campbell, 'The Agrarian Problem'; Campbell, *The Great Transition*, 189.
[110] Campopiano, 'Rural Communities'; Bolòs, 'Changes and Survival,' 328; Galloway, 'Storm Flooding,' 178.
[111] Steinberg, *Acts of God*.

churches, mosques, and other communal places of worship offer a reservoir of social capital which can be employed to organize and support relief and recovery efforts.[112]

If true for present-day societies, then this is perhaps even more apparent for past ones, given that religion was for pre-modern societies a powerful force that pervaded almost every aspect of life. In pre-modern Europe an awareness of the laws of nature was by no means absent, but people were nevertheless inclined to attribute calamitous events such as devastating floods or earthquakes to the wrath of God, turning to prayer and religious rituals for protection.[113] When in 1634 a major flood hit the coast in the North of Germany, for instance, this was attributed to the will of God; the extraordinary dimensions of the flood and the speed with which it struck were considered proof of this. Penance and devotion were deemed necessary for recovery and to prevent a - recurrence.[114] By the early-modern period, most Western European societies 'explained' epidemic diseases through a mixture of frameworks that could co-exist side-by-side: 'miasma' (bad airs and atmospherics), contagion (via people or products), but also providence (from God).[115] Religious institutions also offered practical support in crisis situations – and not just the already-mentioned contribution of poor relief.[116] In Ming China, the impact of religion on resilience worked along different lines. Confucian notions of reciprocity implied that the emperor was responsible for the well-being of his subjects. A failure to see to their basic needs would jeopardize the 'Mandate of Heaven' on which his political authority was based. This provided a strong incentive for the development of the Chinese system of state granaries.[117]

In regions characterized by high-frequency hazardous events, the impact of cultural factors on resilience and vulnerability is closely related to local knowledge and experience. These societies are often well aware of the threat of repetitive and recurring natural hazards such as floods, seismic activity, and droughts. It has been suggested that these 'regions of risk' are often associated with high cultural embeddedness of risk – affecting perceptions and stimulating creative adaptations, and therefore making societies less vulnerable to high-frequency low-amplitude hazards.[118] In contrast, low-frequency high-amplitude events can cause serious disruption even to highly resilient societies because of the

---

[112] Schipper, Merli & Nunn, 'How Religion and Beliefs Influence Perception.'
[113] Gerrard & Petley, 'A Risk Society?'
[114] Jabukowski-Tiessen, "Erschreckliche und unerhörte Wasserflut."
[115] Curtis, 'Preserving the Ordinary.'   [116] See Section 4.3.1.
[117] Brook, *The Troubled Empire*, 109. For the Chinese granary system see Section 5.1.
[118] The expression 'regions of risk' was coined by Hewitt, *Regions of Risk*.

unforeseen nature of the event, and the lack of previous precedent close in time.[119]

Regions of risk then, while confronted with higher levels of hazard exposure, do not necessarily exhibit higher vulnerability: the predictability of a recurrent hazard instead offers the opportunity to learn and to anticipate future hazardous events.[120] This is in line with the IPCC's and disaster studies' dominant 'challenge-and-response approach,' which holds that disasters are triggers for adaptive processes.[121] According to Franz Mauelshagen, all strategies of coping are based on the expectation of repetition drawn from the experience of repeated disasters.[122] A renowned example of such adaptation because of anticipation concerns the 'amphibious cultures' in the coastal plains along the North Sea.[123] Living in regions of risk, where virtually no generation could escape a serious flood event, these communities developed 'landscapes of coping.' After every destructive storm, dikes were raised to prevent a recurrence; settlements were moved to safer places and drainage projects became increasingly sophisticated. Reconstruction efforts were accompanied by technological innovations such as the introduction of wheelbarrows, but also required organizational adaptations: cooperation between landowners was regulated by 'dike laws' that clarified the rights and duties of each and made arrangements for conflict resolution.[124] In the Low Countries, this mindset has been referred to as the 'poldermodel,' whereby collective action and bottom-up decision-making became culturally ingrained as a way of managing complex water management tasks.[125]

Similar societal adaptations are found in other types of risk societies as well. In pre-Hispanic Mexico, for example, the awareness of recurring droughts induced pre-colonial communities to construct irrigation systems, rely on mixed farming, and maintain a safe level of seed and grain in stock, to reduce vulnerability to likely future droughts.[126] In the pre-modern Campine area (Low Countries), the insidious threat of drifting sand was well known to the rural communities. They actively geared up against this hazard by collectively planting windbreaks and enclosures to stop the sand from drifting and, through the local decision-making institutions, they also prohibited the uncovering of the bare soil.[127] This kind

---

[119] Endfield, 'The Resilience and Adaptive Capacity,' 3677.   [120] See also Section 5.2.1.
[121] Noble et al., 'Adaptation Needs and Options.'
[122] Mauelshagen, 'Flood Disasters and Political Culture,' 134.
[123] Van Dam, 'An Amphibious Culture.'
[124] Mauelshagen, 'Flood Disasters and Political Culture,' 133–139.
[125] For a discussion on the existence of the polder model see: Soens, 'Polders zonder poldermodel?'
[126] Endfield, 'The Resilience and Adaptive Capacity,' 3677.
[127] De Keyzer, "All We Are."

of behavior has been described as 'subcultures of coping'; a concept which refers to cultural patterns that – usually out of necessity – are geared towards accommodating problems and risks arising from an awareness of a persistent disaster threat. Risky and unstable environments fostered particular patterns of behavior, social structures, and institutions to build resilience and 'normalize' these hazardous recurring life experiences.[128]

In his work on the Philippines, Bankoff explains how the entire (pre-colonial as well as colonial) culture of the islands was formed by the experience of recurrent seismic and meteorological hazards in this region of risk, in order to reduce the level of vulnerability. Indigenous building techniques, for instance, were adapted to environmental risks as the use of light materials such as nipa palm and bamboo minimized casualties from earthquakes, while low ceilings reduced the damage incurred from typhoons. To cope with adverse circumstances, agricultural systems were geared to ensure food security through crop diversification, land fragmentation, and the use of trees as windbreaks. The 'culture of disaster' in the Philippines also includes the practice of moving out of harm's way by resettling in a safer location, plus a number of cognitive strategies and ideological elements such as a reliance on the 'leave it to fate' sentiment (*bahala na*) as a sense-making strategy, strong group cohesiveness, and exchanging jokes about disastrous events as a way to relieve anxiety and psychological distress.[129]

It would be a mistake, however, to believe that all societies facing recurrent hazards are able to seamlessly adapt, for the development of a culture of coping may be impeded by political, social, or economic circumstances. In Northern Germany, Mauelshagen notes significant land losses due to storms in the late fourteenth century, possibly because, after the Black Death and recurring plague outbreaks, the manpower needed for the repair and construction of dikes was lacking, and again in the seventeenth century, when the Thirty Years War placed a heavy fiscal burden on the population of the coastal region.[130] Likewise, Eleonora Rohland points out that in eighteenth- and nineteenth-century New Orleans, recurrent hurricanes and floods did not give rise to sophisticated infrastructural designs to reduce the level of vulnerability. The French and Spanish colonial authorities prioritized short-term strategic interests and built forts in flood-prone locations, ignoring both traditional environmental knowledge and observations from their own engineers. Relief and reconstruction after the hurricane of 1812 were prevented by

---

[128] Bankoff, 'Cultures of Disaster.'   [129] Bankoff, *Cultures of Disaster*, 163–170.
[130] Mauelshagen, 'Flood Disasters and Political Culture,' 135–136.

the political turmoil and racial issues that had emerged after the purchase of Louisiana by the United States in 1803.[131]

Overall then, alignments of interests and distributions of power in decision-making are crucial determinants of whether cultures of coping are sustained in ways that reduce vulnerability. A mismatch of priorities between what 'outsiders' consider as disaster risks and the different ways in which risks are perceived and responded to by 'insiders' can cause 'culture gaps,' which may lead to negative outcomes for those who have little say.[132] This could be applied to the attempts of 'outsiders' to implement containment measures for Ebola-afflicted communities in Western Africa – with resistance from 'insiders' stemming from greater precedence put upon maintaining traditional and customary practices, social networking, and economic activity.[133] Such patterns have also been noted with respect to contemporary disaster management, where the prevailing disaster risk reduction rationalities do not always align neatly with culture, or where local elites communicate concerns different from those of the majority of the population.[134]

---

[131] Rohland, 'Adapting to Hurricanes,' 6–9.
[132] Krüger *et al.*, *Cultures and Disasters*.
[133] Cohn and Kutalek, 'Historical Parallels.'
[134] Krüger *et al.*, *Cultures and Disasters*.

5 Disaster Responses

While the previous chapter focused on the pre-existing conditions and pressures that helped determine the impact of hazards, this chapter looks at societal responses to hazards – active decisions, policies, and adaptations aimed at enhancing the resilience and reducing the vulnerability of societies or groups. This covers interventions as varied as adaptations of land use or building styles to environmental conditions, physical protection against hazards, the use of warning or prediction systems, measures to contain the spread of disease, or the distribution of disaster relief. The distinction between the initial shock, on the one hand, and its eventual impact, on the other, implies that all such measures can be seen as forms of disaster prevention: even when the shock itself cannot be avoided, it may well be possible to prevent it from developing into a disaster. That is why we refrain here from distinguishing between 'structural' measures that modify the event and 'non-structural' ones that modify human vulnerability.[1] Instead, in this chapter we discuss three important general aspects of responses to shocks: the relationship between top-down and bottom-up initiatives, learning from experience and the application of 'expert' knowledge, and the constraints on active responses to hazards – especially via the unequal application of power and wealth.

5.1 Top-Down and Bottom-Up Responses

Previously we discussed the increasing involvement of the state in disaster management since the early-modern era, and subsequently we pointed to the importance of collective action: formal or informal cooperation and coordination at the local level.[2] In terms of this distinction, responses can be arranged between two extremes. At one end of the scale are top-down interventions imposed by a government or elite, requiring substantial investments and making use of advanced technological knowledge. At

---

[1] For a categorization of this type: Smith & Petley, *Environmental Hazards*, 73.
[2] See Sections 4.3.1 and 4.3.2.

the other are bottom-up initiatives that rely on local resources and know-how. In many cases elements of both are present: that is to say, even top-down initiatives are not entirely 'imposed' but simultaneously require acceptance and implementation at 'street level' too.

While the technological advances of the modern era have stimulated the use of top-down interventions, it would be a mistake to think they were absent earlier. A prominent case is the sophisticated 'ever-normal' granary system established in Qing China, intended to stabilize grain prices and provide a safeguard against famine. Although the system built on some of its predecessors, it was greatly expanded and refined in the late seventeenth and early eighteenth centuries, reaching its greatest complexity in the middle decades of the latter. By then, granaries had been established in all main population centers. In order to keep reserves in these granaries at stable ('ever-normal') levels, they were regularly restocked through purchase, tribute, or private contributions in exchange for titles of honor. Transport – mainly by river or canal – was organized from granaries in regions with a surplus to meet shortages elsewhere in the empire. In years of poor harvests, grain was distributed at reduced prices to the entire population; stocks were also used to minimize seasonal price fluctuations. All of this required very substantial financial resources and bureaucratic efforts by the state.[3] It is estimated that in dearth years granary distribution would feed around 5 percent of the population of any province for two months, while the total costs of the granary system amounted to 0.5 to 2 percent of the state's annual revenues.[4]

In situations where state power was not as strong, top-down intervention might take a somewhat different shape. In the late twelfth century, the urban authorities of Verona, in Northern Italy, tried to prevent a recurrence of the famine of 1178 by actively increasing the area of land under cultivation in the surrounding countryside. Marshlands were granted to wealthy citizens who started reclamation projects, whereas more arid land, after an irrigation canal had been dug, was divided into farms onto which a community of peasant families was settled.[5] In the first case, then, the Chinese authorities simply had the power and resources to coerce implementation of the necessary prevention measures, whereas in the second case, Italian city-states still had to make some concessions to get the process under way. In fact, it was often the case that even when ambitious authorities did initiate top-down preventive measures, they still had to fall back on local knowledge and networks. For

---

[3] Will & Wong, *Nourish the People*, esp. 43–73.
[4] Will & Wong, *Nourish the People*, 481–484, 493–494.
[5] Campopiano, 'The Evolution of the Landscape', 324.

example, in the late-eighteenth-century Southern Low Countries, the government imposed a series of drastic measures to contain recurrent outbreaks of the rinderpest, including the compulsory slaughter of all cattle on farms where the disease was identified (Figure 5.1). The policy was carried out despite resistance by the farmers, but in order to achieve this process of 'stamping out' the authorities had to rely on the ability of experienced local lay veterinary healers present in every community to diagnose the disease.[6]

However, disaster management specialists have argued that while top-down interventions may protect lives and livelihoods, they may also generate or exacerbate vulnerability.[7] The first phase of the Greater Dhaka Flood Protection Project carried out in the 1990s is a good example. This large-scale engineering project, aimed at preventing the flooding of this area of Bangladesh, turned out to have many unexpected

Figure 5.1 Engraving showing the drastic measures imposed by the Dutch government in 1745 to contain the rinderpest – the compulsory slaughter of all cattle on farms where the disease was discovered – and the resisting farmers. Caption: Gods slaandehand over nederland door de pest-siekte onder het rund vee naa het leeven getekent en gegraveert door Jan Smit (1745). Courtesy of the Rijksmuseum, Amsterdam.

---

[6] Van Roosbroeck, 'Experts, experimenten en veepestbestrijding.'
[7] Blaikie et al., At Risk, 218.

side-effects in the areas just outside the embankments. There, the inflow of polluted water forced many residents to abandon agriculture, resulting in a decline of average income. In addition, serious problems emerged with waste disposal, water quality, and sewerage, posing risks for public health.[8]

Accordingly, bottom-up measures may be more attuned to local needs: agricultural systems, settlement location, and construction methods were often adapted to the natural environment in order to minimize risks. In the Campine region of the Southern Low Countries, located in the European sand-belt region, sand drifts have long posed a serious threat to farms, fields, and villages. Yet in the late Middle Ages the inhabitants alleviated ecological degradation through different prevention techniques. A framework of collective resource management and property rights stimulated members to plant shrubs and trees and restrict harmful practices such as peat extraction and mowing. Inhabitants had the capacity to do this, since powerful elites were not present to violate, obstruct, or ignore communal regulations, and inhabitants also had the incentive, since their welfare was intrinsically embedded within the sustainability of the socio-ecological system. However, prevention strategies of this type require a thorough understanding of local conditions. If change happened very quickly or if the area was settled by newcomers from other agricultural regions, preventative measures might not develop. In the early Middle Ages the Campine region did witness the destruction of settlements by sand storms, for newly arrived settlers reclaiming the land were unaware of the risks of deforestation and large open fields in this environment, setting irreversible sand storms in motion.[9]

Epidemic disease was one threat that clearly brought about a series of complex prevention strategies in the pre-modern period – especially those introduced by urban governments. Although contagion was often seen in a context of divine punishment, pre-modern European and Middle Eastern societies did understand contagion – viewed through the lens of miasmatic theory, for instance.[10] Accordingly, plagues and other diseases fashioned the need for urban governments to impose various kinds of practical regulations and restrictions, many of which still feel remarkably familiar to anyone living through present-day outbreaks of epidemic diseases such as the 2019–20 coronavirus disease COVID-19. Some of the earliest plague ordinances can be found in the fourteenth century, though they increased in frequency from the fifteenth century onwards, and some have

---

[8] Alam, Damole & Wickramanayake, 'Effects of Flood Mitigation Measure.'
[9] De Keyzer, "All We Are."
[10] Shefer-Mossensohn, 'Rethinking Historiography,' 15–16.

been described as particularly successful – the rapid response by the Neapolitan government in the 1690s kept plague in Apulia restricted to just ten agro-towns,[11] while Ragusa (Dubrovnik) was already plague-free during the early stages of the sixteenth century – a public health success story.[12] Most ordinances were concerned with the same things, even if they had local variations: (i) isolation of the sick or suspected sick in their houses, (ii) maintaining hygiene in public spaces, (iii) the regulation of trade, and (iv) supervision and social control. In connection with this, some scholars have argued that the fear of epidemic outbreaks was instrumentalized by 'elites' or 'authorities' as an 'excuse' for implementing order.[13] In pre-modern Europe, social control is said to have taken the form of heavy impositions on certain vulnerable groups: beggars and vagrants, prostitutes, foreigners and outsiders, and the general poor. In some cases, straightforward persecution took place, as was the case with the carefully orchestrated pogroms against Jewish populations during outbreaks of plague in the later Middle Ages.[14] As well as offering the blueprint for social behavior, authorities also developed systems to enforce compliance – often through onerous punishments, usually of a financial nature.[15]

As mentioned above, however, top-down authoritarian attempts to implement behavioral patterns that reduced the likelihood of an epidemic disease, or restricted its spread once in motion, could not really work without cooperation from below. Evidence from pre-modern Europe suggests that these kinds of impositions were rarely accepted automatically. In fact, rather than being a suitable vehicle for overly draconian measures and social control, plague epidemics in particular became a context for those lower down the social hierarchy to vent their frustrations and fears, if not in the same politically charged manner as seen in the nineteenth- and twentieth-century cholera outbreaks.[16] More often this was only in a passive way via a refusal to cooperate with the authorities, but on occasion it included more assertive and direct action leading to conflict. Examples of this kind of failure to do the authority's bidding included the rejection of supervised and restricted burial practices. Instead, family, friends, and neighbors continued to carry and transport infected bodies – despite dangers to themselves – and often erupted in violent protest when prevented from doing so.[17] Frequently, we find

---

[11] Fusco, 'The Importance of Prevention.'
[12] Blažina Tomić & Blažina, *Expelling the Plague*.
[13] For the original work: Foucault, *Discipline and Punish*, 195–200. Examples of this idea: Lis & Soly, *Poverty and Capitalism*, 79; Naphy & Spicer, *The Black Death*, 80.
[14] Cohn, 'The Black Death and the Burning of Jews.'
[15] Curtis, 'Preserving the Ordinary.' [16] Cohn, *Epidemics*.
[17] Curtis, 'Preserving the Ordinary.'

people refusing to pay fines for contravening regulations and rejecting orders to move from their homes to isolated quarantines. Given that communal suspicions and distrust of the decisions and actions of 'elites,' 'authorities,' 'experts,' or 'outsiders' are seen even today in the wake of serious epidemics, with the case of the 2013–15 Ebola outbreak in West Africa as an obvious example, leading to violent resistance and outcry at the local level,[18] the need to engage local communities and respect cultural contexts when developing epidemic prevention and containment strategies is something that runs very deep across different societies, but also goes far back in time.

5.2 Experience, Memory, Knowledge, and Experts

5.2.1 Memory and Learning from Experience

The belief that it is possible to learn from disasters is a vital element in the disaster cycle. Memory and learning from past experiences affect both vulnerability and resilience. Anticipating recurrent hazardous events may trigger investments in infrastructural works, changes in societal structures and reorganization of institutions to prevent a future hazard from unfolding into a disaster, thus reducing vulnerability. When eradicating the hazard or shock altogether is not possible, learning and memory can help to mitigate the effects of the shock and enable a much faster recovery or adaptation. This influences resilience as well. Memory and learning can take place on various levels. Governments and other policy makers can draw conclusions from what went wrong – especially in present-day situations, they can frequently avail themselves of the results of official investigations – in order to take technical and managerial measures, but memory and learning are just as much a matter of common people applying practical knowledge in day-to-day-life, individually and collectively.

Reactions to famines in late-twentieth-century Africa provide an example of the former situation, where changes of government policies occurred on the basis of experiences during earlier disasters. In Ethiopia and Sudan national authorities and NGOs drew lessons from the terrible famines of the 1970s and 1980s to prevent a recurrence during the droughts of the mid-1990s. Projects introducing technological innovations in agriculture were initiated, and both countries – Ethiopia in the late 1970s and Sudan in the mid-1980s – established agencies that were given the task of setting up and maintaining early-warning systems and

---

[18] Kutalek et al., 'Ebola Interventions'; Cohn & Kutalek, 'Historical Parallels.'

establishing national cereal reserves. In cooperation with NGOs, food-for-work programs were initiated to provide food rations to hundreds of thousands of households in return for labor, and food aid from abroad increased substantially. Health services and sanitation were improved.[19] Policies were successful in the sense that the threat of famine was addressed more effectively and quickly in 1994 than in earlier decades.[20]

But while resilience has improved, vulnerability remains: famine is still a very real risk in parts of Africa. Although the governments of Ethiopia and Sudan also tried to redirect agricultural policies towards national self-sufficiency, in this respect they were much less successful.[21] In addition, social and political factors affect vulnerability through the marginalization of social or ethnic groups and the recurrent armed conflicts that exacerbate the effects of droughts. In some cases famine is even used as a deliberate strategy in those conflicts. As the authors of a recent report on African famines over the last three decades conclude, learning lessons is one thing, but actually translating them into action is another. Early-warning systems, for instance, have little effect if the warnings are not followed by 'early action.' When learning takes the form of adjustments in government policies, the political will to act upon the insights gained from experience is decisive.[22]

Political will does not carry the same weight in situations where learning is a bottom-up process. A good example is the culture of coping with hurricanes or typhoons in the Philippines, which has already been discussed above.[23] The impact of the recurring typhoons was actively mitigated through preventive building methods and infrastructural works, but also through the mental acceptance of risk and a strong sense of solidarity. Since everybody was susceptible to risk, relief systems and redistributive mechanisms were immediately put in motion to help neighbors and community members who had been struck during a hazardous event. These informal networks were eventually adopted by the Spanish rulers of the islands in the form of fraternities called *cofradías*.[24] Collective action mechanisms and the culture they supported secured continuity in the face of risk and helped prevent most hazards from turning into true disasters.

A disaster memory often requires decades or even centuries to develop and relies upon a certain level of demographic stability. Only when communities can get acquainted with an environment and its hazards can traditional environmental knowledge be formed, which is the basis of

---

[19] Von Braun, *A Policy Agenda*, 14–19.
[20] United Nations Economic Commission for Africa, *A Symposium*, 6.
[21] Von Braun, *A Policy Agenda*, 13–14.
[22] Devereux, Sida & Nelis, *Famine*, esp. 11–14, 27–28.   [23] See Section 4.5.
[24] Bankoff, 'Cultures of Disaster,' 268–273.

a culture of coping. This is not evident in more mobile or rapidly changing societies. In New Orleans, for example, traditional environmental knowledge was continuously lost. The French and English settlers did not inherit or appropriate elements of environmental knowledge from local communities. Strategic interests induced the colonizers to locate their settlement in a vulnerable zone. Most of the city was surrounded by water, and drainage issues were a significant problem. But, on arrival, the settlers were unaware of the recurrent hurricanes. This was demonstrated by the unforeseen and hence disastrous hurricane and inundation in 1722 within four years of the foundation of the city. Unlike other settlements in the Americas that proved to be ill-located, New Orleans was not relocated or deserted for a more suitable site. The plain of Manchac was deemed an 'indispensable necessity' for the new colonial powers. Moreover, in New Orleans successive colonial powers followed each other at relatively high speed. By the second half of the eighteenth century, when local environmental knowledge had been developed, the colony had been given to the Spanish and the process had to start again. Distrust of French settlers and fear of slave revolts in the wake of hurricanes prevented an efficient response to the disasters. As a result a culture of coping did not develop despite the repetitive character and nature of the environmental hazards.[25]

Finally, lessons from the past do not automatically stick forever: experience and knowledge may fade when they are not maintained. Such instances of forgetting or 'de-learning' can be triggered by a reduced frequency of shocks, either because of changes in the natural environment or, ironically, as a result of effective disaster prevention. This may give rise to a false sense of security, similar to the 'levee effect' discussed above,[26] in turn leading to the erosion of the institutional framework required to maintain expertise and ensure its transfer to the next generation. The outbreak of the plague in Surat (Western India) in 1994, for instance, took place almost thirty years after the last case of plague in the region and more than sixty years after the last major epidemic.[27] The outbreak was caused by a combination of factors. These included, among others, unhygienic living conditions in the city's crowded slums, but also the abandonment of plague monitoring schemes due to budget cuts, related in turn to the austerity measures imposed in order to ward off a national debt crisis. These schemes were easy targets in a situation in which the immediate threat of the disease seemed no longer present.[28]

---

[25] Rohland, 'Adapting to Hurricanes'.
[26] For a description of the 'levee effect', see section 4.2.1.
[27] Barrett, 'The 1994 Plague'.   [28] Barnes, 'Social Vulnerability.'

5.2.2   The 'Rule of Experts'

"Don't find a fault. Find a remedy." This quote from Henry Ford included in the 2006 Bipartisan Katrina Report, 'A Failure of Initiative,' reflects how present-day governments and societies aim to deal with disaster.[29] The remedy usually suggested is technological and those suggesting it are usually 'experts,' that is, scientists and engineers. This dependence on experts and their mastering of techniques and technology to deal with and adapt to hazards and disasters is, however, not self-evident and can be better understood by tracing the historical roots of this phenomenon and its evolution throughout time.

The rise of experts as a separate social group with a separate kind of 'expert knowledge' is usually situated at the end of the sixteenth century and the beginning of the seventeenth, and is often considered a crucial feature of the modernist interaction with nature.[30] As argued by Eric Ash, the meaning of the concept 'expertise' in late-sixteenth-century England gradually shifted from practical experience in a certain field to insight in more abstract, mathematical models. When it came to navigation, for example, in the early sixteenth century expertise was seen as belonging to well-versed seamen, who had elaborate experience of finding their way on the ocean. At the end of the same century, navigational expertise was expressed in complex mathematical models, in the new genre of 'navigational handbooks.'[31]

The new, late-sixteenth-century definition of expertise heralded the emergence of a new social group, the 'experts,' who used this new idea of expertise to establish their own position. Their ally in this pursuit was the emerging early-modern state, which legitimized their expertise and used experts for its own gain: as mediators and project coordinators they allowed the state to expand its scope and undertake projects in more distant parts of the realm. Experts and the state thus co-evolved.

In the modern era, the rise of the nation state and the advance of technology and science gave the rule of experts a dominance it had not previously possessed. In the middle of the nineteenth century 'social engineering' and techno-politics flourished, as the new nation states left behind their *laissez-faire* social policies – inspired by the Neo-Malthusian idea that demographic checks were necessary – opting instead for a new focus on proactively developing social measures. This development was not restricted to the Western world: it can also be observed in other parts of the globe colonized by European powers.[32] In older studies the

---

[29] See www.npr.org/documents/2006/feb/katrina/house_report/katrina_report_full.pdf.
[30] Long, *Engineering the Eternal City*.   [31] Ash, *Power, Knowledge, and Expertise*.
[32] For instance in Egypt under British rule: Mitchell, *Rule of Experts*.

increased state intervention and its heavy use of science and technology have often been seen as a story of progress, and of the success of modernity.[33] This is not without reason, for state-sponsored technological innovation based on scientific expertise did sometimes indeed contribute to a reduction of disaster vulnerability. Some of the most spectacular examples can be found in the battle against infectious diseases in European cities in the second half of the nineteenth century. By the 1860s it had become clear that diseases such as cholera were spread by contaminated water and that the existing water wells and sewers were to blame for the spread. It was physician John Snow who discovered in 1854 that cholera was conveyed by water and that the pandemic of London was linked with one very particular pump in Broad Street. The ground water of the well in Broad Street was contaminated by the water seeping from the different cesspools in the area. As a result, urban as well as national governments in Europe and America became convinced that sanitary reform was urgently needed. In England Sir Edwin Chadwick, who led the Poor Law Commission, urged a complete reconstruction of the poor neighborhoods and advised a sanitary reform, whereby sewers and piped water supplies were provided in all English towns for all social classes. Soon, similar measures were taken in towns and cities in other European countries, with very favorable results.[34]

The rule of experts, however, also has a reverse side: states may use their social policies to control the lives and bodies of their subjects – biopolitics as Foucault would have called this. Chadwick, in his famous 1842 report on *Sanitation*, was very explicit in his – in hindsight very Foucauldian – ambitions: implementing centralized sewer systems and piped water networks would mean that more poor people would be healthy and could earn a decent living through working.[35] Nineteenth-century state intervention in other social domains bears witness to a similar approach. Hunger, for example, became a 'social problem,' and practitioners in the emerging fields of nutrition and domestic science helped craft a response to this problem. The introduction of British school meals, offered to those children who failed to develop according to scientific benchmarks, is a clear case in the battle against hunger, influenced by this cooperation between state and science.[36]

There is a vast literature on how the combination of state control and the rule of experts disrupted normal routines in socio-environmental systems, often with devastating results. The irrigation agriculture in the

---

[33] Some poignant examples of this can be found in Shapin, 'Science.'
[34] Baldwin, *Contagion and the State*, 147–156.   [35] Hamlin, *Public Health*.
[36] Vernon, 'The Ethics of Hunger,' 696.

Egyptian Nile Delta offers a famous example of a highly complex socio-environmental system, the maintenance of which was largely devolved to individual village communities, at least from the Mamluk period onwards. From the late eighteenth century onwards massive public dam and canal building – such as the reconstruction of the Ashrafiyya/Mahmudiyya Canal, linking Alexandria to the main Nile branches in 1816–19 – required a forced mobilization of thousands of peasant laborers, not only resulting in a high number of casualties among the workforce, but also disconnecting peasants from their homes and traditional knowledge of their environment. These grand projects would not have been possible without the combination of a new type of ruler, in this case, *khedive* Mehmet Ali, 'the father of modern Egypt' (1805–48), and a new type of expert, personalized by Rohidin, an Ottoman engineer from Istanbul and founder of the Egyptian School of Engineering in 1816.[37] According to Alan Mikhail, these public dam-building projects, from the Mahmudiyya Canal to the Aswan High Dam in the 1950s and 1960s, not only constituted a complete make-over of both the environment and the rural society of the Nile Valley, but also increased the vulnerability to natural hazards, which came in the form of salinization, shrinkage of the Delta, siltation of the irrigation network, and evaporation of water.[38]

The Foucauldian perspective, in which expertise and technology are used by states and their expert-allies to control people, often by controlling nature, also emphasizes how this led to an increased vulnerability among the 'powerless.' Through their impact on the environment, technological systems help to produce and reproduce social inequalities and power relations. In risk-prone regions, the use of capital-intensive technology to counter hazards such as flooding or earthquakes usually privileges the wealthier districts, either directly or indirectly, as the real estate market will push poorer people towards the least protected areas.[39]

What is more, technology in itself may also lead to hazard and disaster, as in the case of 'technological lock-in.' The choice for a specific type of technology creates a certain path-dependency and can create its own vulnerabilities. A clear example of technological lock-in is the case of Dutch flood safety. When the Dutch started draining their wetlands around the year 1000, they set in motion a process that was hard to stop. The land subsided when it had been drained, so dikes (seawalls and river embankments) had to be built. Owing to the lack of flooding, the deposition of sediments decreased, meaning dikes had to be raised even

---

[37] Mikhail, *Nature and Empire*, 260.
[38] Mikhail, *Nature and Empire*, 254–255, 294–295.
[39] For historical examples, see Soens, 'Resilient Societies.' For a case study on Hurricane Katrina, see Tierney, 'Social Inequality.'

more, and so on, and so on. The confidence of the Dutch in this type of technological solution is large – especially after the Delta Works, following the 1953 floods, which kept Dutch feet dry for over sixty years. The chances of failure seem slim (standards are set at once every 10,000 years in densely populated areas and once every 4000 years in less densely populated districts), but this ignores the fact that the calculations of flood safety levels are highly uncertain and that if flooding happened the consequences would be disastrous, as millions of people currently live below sea level. The impact of climate change accelerates the process of falling land levels and a need for rising levees, which in the long run seems untenable. Thus "the Netherlands truly finds itself in a lock-in because there is no question the entire population in low-lying areas can move elsewhere."[40] However, during the last couple of years the 'accommodation paradigm' (i.e. leaving space for water) has been slowly gaining ground, indicating that political and economic choices might indeed provide a way out of technological lock-ins.

5.3   Constraints on Disaster Responses

5.3.1   *Inequalities in Power and Property*

Previously, inequality has been discussed as one of the pre-existing conditions that help determine the impact of hazards.[41] Inequalities in wealth, resources, and power are also among the factors that may constrain the ability of societies to respond. Coping strategies for the 'common good' were sometimes blocked or hindered, if they conflicted with the goals and desires of those with more resources or wielding more power, and on occasion the responses that were allowed to be enacted benefited only a small elite group rather than society at large.[42] It is clear that one society's capacity to respond to a hazard under conditions of plantation slavery was different from another society's response capacity under conditions of feudalism, and this again was different from the range of responses available to people in peasant societies or free market-oriented democracies.

In some cases, inequalities in power and property could represent conditions that enhanced responses to hazards – or at least the speed, intensity, and completeness with which they could be implemented. For example, in medieval Europe powerful manorial lords were often influential in laying out open fields, even if this still required negotiation with

---

[40] Wesselink, 'Flood Safety,' 241–242.   [41] For this discussion, see Section 4.4.
[42] Dennison & Ogilvie, 'Serfdom and Social Capital.'

tenant communities. In an open field both demesne and tenant lands were fragmented, mixed, and distributed in the fields. Such spatial spreading of exposure to the elements was an important feature of risk limitation in medieval agriculture.[43] Likewise, powerful urban lords in Northern Italy incentivized (or coerced) rural communities to maintain water defenses in the light of flood risks. These were occasions when elite goals happened to coincide with broader communal ones: manorial demesnes also benefited from the sub-division of plots, while those water defenses allowed wealthy investors to reclaim more land for capital-intensive farming.[44] The same could be said of the process of *incastellamento* in tenth- and eleventh-century Western Europe. Concentration of inhabitants into fortified villages helped protect ordinary people from the hazards of violence, and yet the seigneurial lords who initiated this benefited from collecting a labor force large enough to colonize new areas, and crystallize and secure territorial power.[45]

More often, however, the application of inequitable power or control over resources led to situations where coercive strategies could be employed to stop adaptive responses that were intended to prevent or mitigate hazards. In medieval and early-modern Leiden, for example, tenants and the town council both had interests in maintaining good-quality public health infrastructure, and the disposal of waste into communal cesspits. This was a system that was still in balance in the sixteenth century, but by the seventeenth century it had been replaced by a system of sewers that simply led to the flushing of waste into the city canals, and was likely responsible for terrible epidemic outbreaks that occurred throughout the 1600s. Residents' needs for adequate hygiene provision – with frequent complaints made to the town council – conflicted with the needs of landlords who wanted to keep their costs down (by not maintaining expensive cesspits), and the needs of landlords in an increasingly inequitable city were privileged and protected by an urban council that wanted to expand its housing base to attract migrant workers for the textile industries.[46]

Colonialism was, almost by definition, characterized by unequal power relations and forms of coercion. How this could result in constraints on the response to hazards is demonstrated in early-twentieth-century Malawi, at the time the British protectorate Nyasaland. Here, from the late nineteenth century onwards, a dual agricultural system developed: in the commercial sector cash crops, intended for export, were produced on

---

[43] Curtis, *Coping with Crisis*, 52.
[44] Curtis & Campopiano, 'Medieval Land Reclamation.'   [45] Curtis, 'The Emergence.'
[46] Van Oosten, 'The Dutch Great Stink.'

large estates established on alienated land, whereas traditional subsistence agriculture was relegated to the increasingly congested but less fertile lands reserved for the native population. By the 1940s many smallholders combined wage labor in the commercial sector or in the emerging industries of South Africa, Southern Rhodesia, and neighboring Northern Rhodesia with subsistence agriculture on their own small plots of land.[47] Increasingly they focused on maize instead of sorghum or cassava. Maize yields a high caloric value per unit of land or labor, but is also very sensitive to water deprivation.[48] As the dependency on maize increased, so did vulnerability to drought. Although the drought-induced famine of 1949 was a minor one in numbers of victims, the fact that it occurred nevertheless demonstrates the narrowing of traditional coping mechanisms under colonial rule. Moreover, the consequences were long-lasting: in combination with other factors, the lack of diversification in agriculture gave rise to another famine in 2001–02.[49]

Although low levels of inequality do not guarantee an adequate response to hazards, they do offer advantages. In more equitable societies both assets and risks are more evenly distributed. This balancing removes an important source of conflict when it comes to responding to hazards, for it is more likely that responses that benefit elites also benefit others. Moreover, in economically equitable societies political power tends to be shared by a larger group: these societies are usually characterized by collective-choice arrangements that impose restrictions on short-term rent seeking and prioritize sustainable solutions. With their emphasis on local, bottom-up governance and rules design, they also promote flexibility and adaptability.

Flexibility or the opportunity to adapt is a recurring trope in the literature on the resilience of communities. The idea that societies where resources and power are more equally divided are better at adapting is often implicitly present within the literature on 'institutional resilience,' with its emphasis on local, bottom-up governance and rules design.[50] Likewise, anthropologists and historians have pointed to the fact that institutions are not necessarily efficient and are often very much steered by those in power. Ensminger, for example, focusing on the Orma people in twentieth-century Kenya, pointed to the fact that new institutions evolved very much according to the needs of those with the highest bargaining power and most resources.[51] Similarly, work on the Kafue floodplains in Zambia showed how, in the transition from colonial rule to

---

[47] Vaughan, 'Famine Analysis.' [48] McCann, 'Maize and Grace,' 249.
[49] Devereux, 'The Malawi Famine.' [50] See also Section 4.3.1.
[51] Ensminger, *Making a Market*.

independence, the former local common pool institutions were eroded and altered by being incorporated into a new state structure. Yet, tribal leaders and absentee herd owners were able to steer the new entitlement rules in such a way as to benefit their particular interests, for instance by imposing open access rules without any limit on their use of the common, while disregarding the social and environmental costs of these choices.[52]

By implication, the power balance within a community and the position of the elite relative to other groups – which is linked to economic (in) equality, but in a more nuanced way – played a part in securing or hindering resilience. This is noticeable, for example, in the way poor relief strengthened resilience in different types of communities during periods of grain crisis in the pre-industrial Low Countries. Poor relief played an important part in helping the poorest cope in two different peasant regions. In Inland Flanders, where small, proto-industrial peasants lived in a co-dependency with large tenant farmers, the local elite (that is, these large farmers) decided to invest in relief as it was clearly in their interest to maintain this system of reciprocal – though unequal – exchange, where peasants traded their labor for grain and credit. Their efforts, however, were limited: the main aim was to enhance the status of the donors and maintain a labor force sufficient to meet demand.[53] In the Campine area, a communal peasant-dominated region, where independent peasants governed the villages, a more elaborate and inclusive relief system was in place, as vulnerability was much more equally divided among the people of this region and even local elites were not protected from the risks imposed by natural hazards or infirmity.[54]

The role of elites is also stressed by Di Tullio in the context of resilience to warfare for fifteenth- and sixteenth-century Lombardy in Italy. The Geradadda communities were able to reproduce their social network and contain inequality, thanks to their communal assets, the credit-related opportunities provided by lay confraternities, and the role the elite played in enabling this. Di Tullio sees cooperation among different social groups as quintessential and claims that "reciprocity came before equality"; collective assets were essential to all social groups.[55] The suggestion that we have to look beyond mere economic equality has also been made by de Keyzer in her work on sand drifts in the pre-industrial Campine area in the Low Countries. By combining historical source material with OSL dating, she found that the late-medieval Campine

---

[52] Haller & Chabwela, 'Managing Common Pool Resources.'
[53] For the eighteenth century, see Lambrecht, 'Reciprocal Exchange'; Vanhaute & Lambrecht, 'Famine.'
[54] Van Onacker & Masure, 'Unity in Diversity.'
[55] Di Tullio, 'Cooperating in Time of Crisis.'

village societies were entirely capable of mitigating the effects of sand drifts on the common heath lands that were essential to the economic viability of this region. This resilience can be explained by the fact that all interest groups there (small peasants, village elites, tenant farmers, and lords) relied to some extent on the survival of these commons, thus creating firm incentives for protecting them. Their strong property rights and the powerful grip of the village community on the government of these commons were essential as well.[56]

Institutional resilience – that is, the ability to adapt institutions to changing circumstances – therefore has a nuanced relationship with equality. The literature points firmly to the importance of bottom-up control and collective arrangements, but whether this constellation worked was strongly dependent on the socio-economic context. It is noticeable that a balance of power and shared interests between the social elite and other social groups seem to have been essential in creating flexible institutions that strengthened resilience.

5.3.2 Institutional Rigidity and Path Dependency

Institutional rigidity and the entrenchment of norms, practices, beliefs, and values that shape vulnerability, adaptation, and resilience can act as constraints on responses to hazards – in some cases even pre-determining their outcomes.[57] A common cause of rigidity is path-dependent processes. These processes typically take root during the 'critical junctures' or key moments in time at which institutions are (trans)formed, after which certain directions of change are established and others are foreclosed in a way that shapes developments over long time spans.[58] Importantly for hazard and disaster response, as institutional arrangements become 'locked in' they become much more difficult to change – in a similar way to technological lock-ins – even if they make people more vulnerable and lead to negative disaster management outcomes.[59]

A classic example of institutional rigidity and path dependency is found in former colonial contexts. In these settings, the imposition of colonial rule upon indigenous governance structures was a common critical juncture at which institutions were typically transplanted from the colonizer to the colonized, invariably acting in the interests of the former. Upon independence, however, many post-colonial nations did not simply disband their colonial institutions, but rather these became the new

---

[56] De Keyzer, "All We Are." [57] See also Sections 2.3.5 and 4.3.1.
[58] Mahoney, 'Path Dependence.'
[59] Adamson, Rohland & Hannaford, 'Re-thinking the Present.'

5.3 Constraints on Disaster Responses

apparatus of the post-colonial state, with inherent colonial legacies in their functioning and even in ideology. Macro- and micro-level studies on Latin America have thus shown that the persistence of deep-rooted inequality embedded within colonial institutions constrains the use of resources and thus perpetuates unequal developmental outcomes in the present.[60] Just as many post-colonial nations had continuities from the colonial past in their dominant forms of economic activity, then, it follows that there are also legacies in responses to threats to those activities – as has been shown in governmental drought and famine relief intervention methods in sub-Saharan African nations,[61] and in hurricane response in the Southern United States.[62]

Historians have also shown that institutional rigidity may be influenced by cultural and religious perceptions as much as by political responses – or may even be determined by a hybrid of the two. In pre-colonial societies across much of sub-Saharan Africa, for example, the coming of rain was linked to helpful intervention by ancestral spirits in the heavens – a belief system which spurred 'rainmaking' rituals in times of drought. As some of these societies evolved into more centralized state structures (a pre-colonial critical juncture), links between rainfall and ancestral spirits became tied to the political leadership, who were believed to have the ability to intercede directly with the heavens and bring rain.[63] Ultimately, the entrenchment of these links between institutions and the environment could – and, according to some, did – have negative consequences for the fate of the political leadership or even state structures as a whole in times of protracted environmental stress, or could lead to blame and scapegoating of marginalized or minority groups.[64]

In turn, colonial actors brought their own perceptions and usually attempted to marginalize indigenous belief systems, although the extent to which they succeeded in doing so was context-dependent. In the Southern African case, even during the seventeenth-century high water mark of Portuguese power, indigenous beliefs and perceptions concerning rainfall were disparaged but did not disappear – an outcome that likely had as much to do with the weaknesses of the Portuguese as with the deep-rooted nature of indigenous norms that had built up over several centuries. In other cases, however, we observe that it was the perceptions of the colonists that were remarkable for their change rather than their rigidity. For example, in Jamaica it has been shown that while the English

---

[60] Mahoney, *The Legacies of Liberalism*; Mahoney, *Colonialism and Postcolonial Development*.
[61] Devereux (ed.), *The New Famines*.   [62] Rohland, *Changes in the Air*.
[63] Hannaford & Nash, 'Climate, History, Society.'
[64] Huffman, 'Climate Change during the Iron Age'; Brook, *The Troubled Empire*; Klein et al., 'Climate, Conflict and Society.' See also Sections 4.5 and 6.1.4.

imported their puritan 'wrath of God' perceptions of extreme weather events to the Caribbean, this religious perception waned over time as new environmental knowledge was acquired through repeated encounters with hurricanes.[65] That is to say, environmental hazards themselves acted as a partial trigger for (informal) institutional adaptation rather than rigidity, creating new constraints and opportunities for responses to hazards.

The above paragraphs therefore show a double perspective on institutional rigidity, change, and path dependency. On the one hand, political and cultural factors seemingly unrelated to hazards had significant effects on the ability of institutions and societies to cope with such hazards through enforcing rigidity or enacting change. Yet on the other, recurring extreme events themselves could also act to influence institutional change and to overcome rigidities. A challenge for research into institutional rigidity and path dependency, therefore, is to identify potential patterns in its possible causes – in other words: what are the mechanisms by which path-dependent processes come into being or are broken? As we argue throughout this book, this necessitates a move away from the event-based focus of many studies on historical disasters.

[65] Mulcahy, *Hurricanes and Society*.

6 Effects of Disasters

The previous chapters have focused on the preconditions and pressures that make hazards more, or less, likely to occur, and how societies respond to these hazards – with a view to stopping them turning into disasters. It is clear, however, that even in the face of adaptive measures, many societies throughout history could not prevent disastrous consequences – and it is those consequences that are the focus of this chapter. We divide our discussion between those effects seen only in the immediate aftermath or the short term – mortality and victims, demographic recovery, blame, scapegoating, and social dislocation – and longer-term structural consequences for economic reconstruction, social relations, and redistribution of resources. Across the course of the chapter we show that disasters – even ones of similar type and magnitude – did not always produce homogeneous outcomes, and in some cases even led to divergent paths of development in the long term. Furthermore, rather than being totally damaging or even controversially regarded as a 'force for good,' we show that the effects of disasters are best assessed when making a basic distinction between developments at the aggregate level (for example, on the basis of GDP recovery) and those at the distributive level – where it is clear disasters could also be instrumentalized to benefit a certain segment of a given population over others.

6.1 Short-Term Effects

6.1.1 Victims, Selective Mortality, and Population Recovery

Mortality is one obvious short-term consequence of hazards and shocks, although it could also have very significant knock-on consequences in the long term, depending on the scale and nature of death. Mortality has for a long time been a common indicator for measuring the impact of disasters – both for contemporary and for historical societies – but, as we come to see in this section, we still lack solid empirical information on mortality characteristics from many disasters. To some extent, this is down to source limitations. Before the nineteenth century, and even in large parts of the world today, the victims are not always registered meticulously or accurately. This is sometimes exacerbated because certain groups are under-represented in statistics as a consequence of their isolated or marginal position in society. Even when deaths are recorded, problems can emerge: for example, the registration of casualties in selective environments such as refugee camps.[1] Before enumeration by census became widespread in the nineteenth century, church burial records offer insights into excess mortality – although even with this source, coverage is patchy over time and space (and mainly for Europe only), with limitations such as our inability to calculate accurate death rates (percentage mortality as proportion of resident population) and very unsystematic attempts to say much about cause of death.

Moreover, the discussion on mortality as a measure of disaster impact raises even more fundamental issues. One pertinent question is whether mortality figures can actually be a good indicator for the success or failure of a society to deal with a particular hazard or shock. On the one hand, mortality seems like the perfect indicator, as it signals the ability of a society's institutions, technology, and knowledge to offer protection to inhabitants' lives in the face of pre-existing vulnerabilities.[2] However, at the same time, mortality can be a flawed indicator in this regard. First, it often happens that hazards and the disasters that ensue do not kill, but rather destroy capital, disrupt societies and institutions, or ruin ecosystems. When mortality is considered the main indicator, all of these disasters are overlooked or underestimated. Furthermore, some of the health implications connected to disasters play out only over an extended period of time. The famine of the Great Leap Forward in China in 1958–61, for example, created early-life stresses that caused lasting damage to health much further down the line. Individuals born during the famine were, having reached adulthood, more susceptible to hypertension than the non-exposed, and they did not attain the same height. They also worked, on average, fewer hours and their incomes were lower.[3] But second, and more importantly, the causes of death can be unrelated to a resilience or vulnerability framework. In pre-industrial societies in particular, many people died of diseases that were perhaps more likely to occur during periods of hardship, but also could quite reasonably be the result of a random association or chain of events. More fundamentally, mortality

---

[1] For an interesting and unusual early example seen during the disaster relief program of the Yellow River Floods of 1935 in Shandong, China, see Li, 'Life and Death.'
[2] For a positive view on mortality as an outcome variable: Sen, 'Mortality.'
[3] Chen & Zhou, 'The Long-Term Health and Economic Consequences'; Huang et al., 'Early Life Exposure.'

shocks more frequently occurred in what some people may deem a 'successful' society. In the countryside of early-modern England it was often the case that the poorer marginal areas with low levels of aggregate wealth were much more salubrious than those with more commercialized forms of agricultural production – one clear reason being that overall poverty acted as a stimulant to outward migration rather than inward.[4]

Selective Mortality Hazards and shocks kill varying numbers of people, but equally important is that the 'type' or 'status' of those who die is not always the same. Floods are generally not the biggest killers, but of those that do kill, there is often a social profile to the victims – we see this even in modern times with Hurricane Katrina, along the lines of race and socio-economic status.[5] For the pre-modern era, it has been shown in a variety of contexts around the world that floods tended to afflict the poor to a larger degree, because of the greater likelihood that they were residing in places much more unsuitable for human habitation.[6] Indeed, recent literature on the North Sea coastal floods has shown that landless agricultural laborers, often living in flimsy structures close to dikes, were much more likely to die, while wealthy farmers survived.[7] Certain disasters in history did kill large numbers of high-status individuals, however: in Catania during the earthquake of 1693, 62 percent of the clergy were reported to have died. They fell victim to the collapse of many of the city's churches, where at the exact moment of the earthquake an important religious ceremony was going on.[8]

Given that the poor live closer to the edge of subsistence than wealthier groups, it is unsurprising that famines tend to discriminate in their mortality effects according to wealth.[9] The belief that poverty reduces access to scarce food supplies is in fact one of the pillars of Amartya Sen's 'entitlement theory.' Entitlements can be based on the ability to purchase food, but also on production: Sen has shown that, during the 1974 Bangladesh famine, mortality was far higher among wage laborers than among farmers.[10] During the 1984–85 Darfur famine, however, the relationship between economic situation and mortality was largely absent, probably because the effects of occupation were greatly overshadowed by the highly localized impact of disease.[11] In pre-industrial

---

[4] Dobson, 'Contours of Death,' 88–89.
[5] Elliott & Pais, 'Race, Class, and Hurricane Katrina.'
[6] Morera, 'Environmental Change,' 92–93; Perdue, 'Official Goals,' 762; Borsch, 'Environment and Population'; Ali, 'Malign Growth?,' 124.
[7] Soens, 'Resilient Societies,' 163–164.
[8] Condorelli, "*U tirrimotu ranni*," 231, 403–405.
[9] Dyson & Ó Gráda, 'Introduction,' 14–15.   [10] Sen, *Poverty and Famines*, 144.
[11] De Waal, 'Famine Mortality,' 16.

Europe, the extent of famine mortality tended to diverge between urban and rural environments. It was often higher in the countryside, even though that was where food was produced, while the cities stockpiled provisions for their own residents (attempting to avoid social disorder), called upon distant trade links for emergency provisions, and generally had a stronger set of relief institutions.[12] It is no surprise, then, that, during times of famine, country dwellers often migrated to the cities in search of food and resources – a migration process that in turn helped raise mortality in the cities.[13] Famine mortality could also differ between rural environments. For example, in sixteenth-century Italy, mountain communities showed lower mortality rates and recovered more quickly than the lowlands, despite the harsh environmental conditions, due to their lower population densities, isolated location, and diversified production methods.[14] In general, highly specialized regions, especially those focused on grain production, were more likely to witness high mortality rates than regions that diversified their economic activities and grew different crops.[15]

Famines can kill men and women, young and old, at different rates. Significant amounts of research have suggested over the years that women have a greater capacity to survive famines than men and that the underlying cause is their superior ability to deal with periods of acute malnutrition.[16] Various social factors have been posited as causes, including things such as women's 'marketable value' in prostitution, control over household resources, restricted fertility limiting the dangers of childbirth, preferential welfare schemes, and a reduced tendency to migrate, but the overriding explanation is still based on physiology – the 'body fat hypothesis.'[17] Notably, more body fat also means higher amounts of leptin – a key driver of the body's immune system – which is important given that (a) famines often led to death via diseases rather than outright starvation, and (b) modern laboratory work tends to show that adult women are more resistant than adult men to most kinds of bacteria, viruses, parasites, and fungi (with only a few exceptions such as malaria and measles). However, while case study evidence is plentiful for the period after 1850,[18] direct quantifiable evidence for the pre-industrial period is scanty, and we should be careful not to assume that the pre-

---

[12] Curtis & Dijkman, 'The Escape from Famine,' 239–240.
[13] Alfani, 'The Famine of the 1590s,' 31–32; Landers, 'Mortality,' 356–361.
[14] Alfani, *Calamities and the Economy*, 136–168.
[15] For example in the Beauvais region in early-modern France: Goubert, *Beauvais et le Beauvaisis*, 79–80. On the impact of diversification see also Section 4.2.2.
[16] Ó Gráda, *Famine*, 99–102.
[17] With nuances: Speakman, 'Sex- and Age-Related Mortality Profiles,' 823.
[18] Zarulli et al., 'Women Live Longer.'

industrial experience inevitably mirrors the modern. When bioarchaeological or documentary material has appeared, it does not necessarily provide clear proof – at least not to the same extent.[19] Even if we accept the scientific principle that women have 'natural' advantages, certain social conditions can conspire to push women into close proximity with points of contagion or vectors, or limit their access to welfare and resources.[20] Male prioritization during scarcity occurred in Northern India in the early nineteenth century, for example.[21] In terms of intra-household resource distribution during famines, gender-based inequalities also often differed along age lines. In late-Qing China, for example, it has been shown that elderly women were high up in the food hierarchy, because of an enduring principle of filial piety, while younger females and girls were at the bottom.[22]

Moreover, if there was a 'female mortality advantage' during periods of acute malnutrition, superior survival capacities often came together with terrible experiences for women – which links back to our previous statement that mortality is not without its limitations as a disaster effects indicator. In both nineteenth- and twentieth-century famines in China, it has been noted that women could often obtain lifesaving resources such as rice and grain, but at the expense of exploitation – frequently in the form of rape and sexual abuse.[23] Women – especially in the countryside – could also often find themselves abandoned in isolated areas during famines, as more mobile men went off in search of resources or employment in cities.[24] Other works have noted the potential for young girls during food crises to be forced into marriage as minors to obtain the household's early access to marital dowries,[25] and women could be 'pawned' into other households as seen in nineteenth-century Kenya.[26] Posing moral dilemmas on food sharing, these kinds of tactics can be seen in the same light as infanticide – the sacrifice of the individual for a perceived 'greater good.'[27] Linked to this, we should also note that, in certain parts of the world, female children were more likely to be abandoned than their male counterparts in times of hardship caused by

---

[19] Healey, 'Famine and the Female Mortality Advantage,' 186; Yaussy, DeWitte & Redfern, 'Frailty and Famine.'
[20] Curtis & Han, 'The Female Mortality Advantage.'
[21] Sharma, *Famine*, 112. On this concept of male prioritization during disasters more generally: Rivers, 'Women and Children Last.'
[22] Edgerton-Tarpley, 'Family and Gender in Famine,' 142; Edgerton-Tarpley, *Tears from Iron*, 165, 188.
[23] Yang & Cao, 'Cadres, Grain and Sexual Abuse.'
[24] Vaughan, 'Famine Analysis,' 186–189.   [25] Devereux, 'Goats before Ploughs,' 56.
[26] Jackson Jr., 'The Family Entity,' 205–208.
[27] Sen, *Poverty and Famines*, 29; Agarwal, 'Gender Relations and Food Security,' 192.

famine.[28] Certain institutional configurations have been cited as more likely to provide protection to women in times of food crisis and famine, however. The development economics literature has exalted the effects of the commons for women's adaptive capacities – often presented in opposition to the market, which is apparently more likely to be organized by males.[29] Furthermore, an alternative to the 'wife and daughter selling' narrative during famines in Qing China has recently been posed – noting also how, in times of crisis, women would bring an 'extra' husband into the family for support.[30]

The epidemic diseases that were such an important cause of death in the pre-industrial period could also vary in terms of the population sector they targeted, and this is important because it had knock-on consequences for the organization of societies and economies. If more adults of working age died during an epidemic, this created new kinds of societal vulnerabilities – who was left to care for the elderly, the infirm, the disabled, or minors? In underdeveloped and developing countries today, unexpected deaths in adulthood, often during epidemic outbreaks, create significant social problems – orphaned children and uncared for elderly parents,[31] young survivors left with reduced family or community support, weak social networks, and poor access to food and healthcare, and all the while minors assume new roles as 'heads of household,' principal care-givers, and earners.[32] Conversely, if an epidemic killed more of the 'frail' – those seen to be a drain on resources rather than a contributor – it could thereby alleviate a number of societal pressures: fewer to care for, fewer to feed. On a related point, if an epidemic was highly selective by socio-economic status, mainly victimizing the poor (perhaps in specific neighborhoods), this led to little structural change in the economy as new poor migrants simply came in to replace the old ones – a feature often seen in urbanized areas of early-modern Europe.[33] On the other hand, if a disease had 'universal characteristics,' killing a wider range of people, this could destroy a pre-industrial society's human capital levels – and then it was not always a given that this would be quickly and easily replenished.[34] Although diseases such as the plague

---

[28] Strengthening the 'missing women' phenomenon in places such as China and India: Das Gupta & Shuzhuo, 'Gender Bias,' 487.

[29] Agarwal, *A Field of One's Own*, 455. And for eighteenth-century Western Europe it was claimed that the "resources of the commons were often all that stood between [women] and total destitution": Hunt, *Women in Eighteenth-Century Europe*, 148.

[30] Sommer, *Polyandry*, 33, 57–58, 71.

[31] Atrash, 'Parents' Death and Its Implications for Child Survival.'

[32] Ronsmans et al., 'Effect of Parents' Death on Child Survival'; De Vreyer & Nilsson, 'When Solidarity Fails.'

[33] Cipolla, 'The Plague.'   [34] Alfani & Murphy, 'Plague and Lethal Epidemics,' 335.

are often suggested to be selective against the poor – a link posited from late Antiquity to present-day outbreaks in Madagascar – episodes such as the plague of 1629–30 have been shown to be indiscriminate killers.[35]

Population Recovery One difficulty in accounting for the casualties of disasters is the blurred line between immediate mortality and other facets of post-disaster population recovery – connected to fertility, nuptiality, and migration, for example. Given that we do not always have good time series for mortality over long periods, some scholars are reliant on sources giving population counts at different intervals – sometimes with long temporal gaps – and naturally this kind of information tends to hide the reality behind what is driving differential rates of recovery after a disaster.[36]

Some scholars have suggested that population recovery after disasters is influenced less by the immediate death rate, and more by other demographic factors linked to nuptiality and fertility.[37] During serious economic crises – especially those linked to harvest failures and famines – societies have often experienced lower fertility rates, and this could last for some time afterwards – as seen with the famine in China associated with the Great Leap Forward.[38] Especially in the pre-industrial period, marriages were often dependent on the ability to set up an independent household. This required a plot of land, employment, or adequate savings. Disastrous famines, therefore, forced individual households to postpone marriages of their members, and thus birth rates could drop significantly in the first years or even decades after a disaster. Indeed, it has been said that, in many contexts throughout history, there has been a cultural aversion to marriage during times of scarcity, and in any case, even disregarding new marriages, fertility rates went down in existing marriages simply through conditions much less suitable for conception and for successful full-term pregnancy.[39] Parallels can be found in the modern era: in Ethiopia conceptions declined during the prolonged and multidimensional crisis of the 1980s, when a combination of civil war, repression, inflation, and especially famine induced married couples to practice birth control.[40]

---

[35] On plague and poverty: Alfani & Murphy, 'Plague and Lethal Epidemics,' 326; Carmichael, *Plague and the Poor*, 1; Campbell, *The Great Transition*, 306–307. On the 1629–30 plague as universal killer: Alfani, 'Plague.'

[36] See the critique offered in Roosen & Curtis, 'The "Light Touch,"' 36.

[37] This is a distinctive view attributed most famously to the Cambridge Population Group: Wrigley & Schofield, *The Population History*; perhaps in contrast to those more inclined to highlight mortality itself: Hatcher, *Plague*; Benedictow, 'New Perspectives.'

[38] Zhao & Reimondos, 'The Demography.'   [39] Alfani, 'The Famine of the 1590s.'

[40] Lindstrom & Berhanu, 'The Impact.'

This was not always the case for all disasters, however – especially not for those linked to significant mortality spikes such as epidemics occurring outside famine periods (such as plagues). Here, instead, marriages and birth rates may have declined temporarily during the chaos of the peak mortality periods, but in the immediate aftermath, new marriages spiked – especially driven by remarriages of those who had lost their partner to disease.[41] Of course, this depended on institutional factors too – some newly single widows or widowers became attractive to prospective new partners on the basis of inheriting resources from their previous marriage, and in some places, such as Southern Italy, the remarriage of men was acceptable but not that of women (who remained lifelong widows after their husband's death) – making it necessary for some to look outside their immediate localities of residence for new partners.[42] Of those women who could technically remarry, not all decided to do so – those inheriting resources from deceased partners to support themselves independently may have chosen to enjoy freedoms outside the constraints of marriage.[43] Furthermore, in some rural areas men left women and children behind, and did not return – creating sex-skewed habitation patterns in their places of origin, with distorting effects upon marriage opportunities. Accordingly, the rates of remarriage depended on a complex set of pre-existing configurations in inheritance practice, dowry demands, and access to and control of property.[44] This process was often further complicated by the fact that mortality crises caused by famines or epidemics affected not only nuptiality rates, but also the average ages at which people decided to marry, and this had obvious knock-on effects for fertility and the rate at which populations could replenish themselves.[45]

Population levels could also be affected by societal shifts that were initiated during or after a disaster. For example, during the 'calamitous' fourteenth century, many people succumbed to disease, famine, and warfare, but population recovery was sometimes halted, and this did not always have purely demographic causes. Regions that were previously densely populated and supporting labor-intensive arable agriculture in Europe, for example, were transformed into labor-extensive pastoral societies. Before the Black Death the main problem was keen competition for agricultural opportunities, but after the plague the main problem became that there were too few agricultural opportunities altogether – necessitating outward migration to the cities – particularly

[41] Livi Bacci, *La société italienne*, 67–69.   [42] Da Molin, 'Family Forms,' 520.
[43] Franklin, 'Peasant Widows' "Liberation."'
[44] Guinnane & Ogilvie, 'A Two-Tiered Demographic System.'
[45] Carmichael *et al.*, 'The European Marriage Pattern,' 16.

for women.⁴⁶ This kind of rural–urban movement explains why cities in some areas appeared to be more resilient in the face of mortality spikes than the countryside. Not everyone migrated, however: the capacity for people to stay in, or repopulate, a particular area depended on the economic opportunities that were provided for wage labor, access to property, and institutional forms of welfare such as the commons and poor relief.⁴⁷ The same can be seen in modern times. After the major tsunami in 2004 caused by the Indian Ocean earthquake, many Sri Lankan coastal communities were unable to replenish themselves. This was not necessarily only connected with the casualties of the flood wave itself, but also with the inland relocation of the fishing communities after the disaster. Because of government interventions, coastal villages were demolished and replaced by hotel complexes and tourism, which turned the densely populated coastal zones of Sri Lanka into vast uninhabited stretches of beach.⁴⁸

Population recovery was often highly influenced by migration, which is one of the most common responses during and after disasters of all kinds. During epidemics, a commonly suggested pattern was flight from the cities while the disease was active in its worst phases, and rural–urban migration in the aftermath as people looked to fill vacancies.⁴⁹ Epidemics often occurred during periods of warfare – conflicts that frequently bore heavier consequences for rural communities – and thus may have further heightened movement towards 'safe harbors' in the city.⁵⁰ This inward movement of people to the cities is said to be one of the reasons for the establishment of 'urban graveyards' – high-mortality urban demographic regimes – in Europe by the early-modern period and thus shows the circular nature of this process: hazards create movement of people, movement of people creates new hazards.⁵¹ However, it must be noted that quantitative empirical evidence to flesh out these arguments still remains scarce – they are often based on abstract modeling, logical intuition, anecdotal evidence, or evidence of a low-resolution macro nature – or the logic behind the mechanisms posited remains conflicted. Elsewhere, during other disasters, many families may have migrated intending an immediate return to their place of origin, but then were prevented from doing so. For example, after the Lisbon earthquake, several previously poor neighborhoods were not rebuilt in the same way, but the land was

---

⁴⁶ Voigtländer & Voth, 'How the West "Invented" Fertility Restriction.'
⁴⁷ Curtis, 'The Impact.'   ⁴⁸ Klein, *The Shock Doctrine*, 8.
⁴⁹ Borsch & Sabraa, 'Refugees of the Black Death.'
⁵⁰ The 'safe harbors' concept is developed in Dincecco & Onorato, 'Military Conflict'; also Rosenthal & Wong, *Before and beyond Divergence*, 104–105.
⁵¹ Voigtländer & Voth, 'The Three Horsemen.'

used for more prestigious building projects. Accordingly, the rates of replenishment were quite different – not only across localities, but also across neighborhoods of the same locality.[52] In other cases, the rebuilding process evolved more freely – with deliberate incentives such as the reduction of mortgages on sold buildings and plots to entice inhabitants back – as seen in depopulated Catania after the 1693 earthquake that hit wide areas of eastern Sicily.[53]

6.1.2 Land Loss and Capital Destruction

The second-most-important measure of a disaster is capital destruction and how much land is lost or affected. Images of eroding cliffs, mudslides, and inundated coastal estates are often used to show the effects of global warming, as are interactive maps of coastal zones that will be flooded if there is a rise in sea level. This indicator is important not only because loss of land is often traumatic for the affected communities, but also because many disasters cause only minimal human casualties but large amounts of physical destruction.[54]

A case in point is the American Dust Bowl in the 1930s. Although it caused no direct casualties, it has been called the worst human-made environmental disaster the United States has ever experienced. As discussed earlier, the Dust Bowl refers to a decade of extreme soil erosion on the American Plains, stretching from Mexico across the continental United States towards Canada. Owing to the combination of a prolonged and severe drought and the destructive nature of monoculture on the fragile plains, the vegetation cover was reduced fundamentally, allowing wind to sweep away the topsoil. This phenomenon created large dust storms that caused 'apocalyptic' darkness during daylight hours and buried houses, roads, and fields with sand dunes. The worst problem, however, was the cumulative loss of the thin layer of productive topsoil. By the 1940s up to 75 percent of that topsoil had been lost in the most severely affected zones, resulting in large permanent declines in land values – somewhere between 17 and 30 percent per acre, depending on the scale of erosion. The total agricultural loss amounted to 2.4 billion dollars (equivalent to 30 billion in 2007 dollars). Migration was the only option for the most affected farmers and tenants, who moved *en masse* from the plains towards economically more viable states such as California.[55]

[52] Pereira, 'The Opportunity of a Disaster,' 487–488.
[53] Condorelli, 'The Reconstruction of Catania,' 802.   [54] See also Section 2.1.
[55] Hornbeck, 'The Enduring Impact,' 1478. For the literature on the Dust Bowl, see also Sections 1.2 and 4.1.

Figure 6.1 Dust Bowl farm in the Coldwater District, north of Dalhart, Texas, June 1938. Dorothea Lange/Farm Security Administration via Library of Congress.

Although a modernist approach to land exploitation had contributed to the American Dust Bowl, this type of catastrophic land loss due to soil erosion and sand drifts is not simply a modern phenomenon. For example, during the ninth and tenth centuries, the village of Kootwijk in the Netherlands was buried in a dune several meters deep, forcing the community to abandon the site. This medieval dust bowl was most probably caused by large-scale land clearances and the creation of open fields – in sharp contrast to former land use – incorporating smaller dispersed farmsteads within woodlands.[56] Up to today, the active dune has not been stabilized. Even swifter than sand drifts was land loss caused by floods. Again in the Low Countries, river floods often caused havoc, including destruction of mills, sluices, ovens, and other capital goods, as well as the loss of livestock, but not permanent land loss since the water usually receded again after a few months.[57] The fertility of the river clay soils and river valleys often made it

[56] Heidinga, 'The Birth of a Desert.'
[57] Van Bavel, Curtis & Soens, 'Economic Inequality.'

worthwhile to reclaim and embank the land again – a sharp contrast to what could be done in less fertile sandy areas hit by sand drifts. However, permanent land losses were more frequent in the case of sea floods caused by storm surges, tropical hurricanes, or tsunamis. Yet, on balance, recovery of land was still the rule, and permanent losses the exception.[58]

Although warfare may have been the most significant destroyer of capital in the past,[59] arguably the greatest form of capital destruction by specifically nature-induced disasters has been caused by earthquakes – particularly for urbanized areas. The Great Kantō Earthquake of 1923, for example, led to an estimated 204 billion US dollars' worth of damage (2010 HNDECI adjusted).[60] In the course of time, as societies became wealthier and fixed capital goods costlier and more sophisticated, the possible absolute costs of destruction increased in parallel, and in recent decades these have risen faster than the losses of lives.[61] A prime example is the earthquake, and the ensuing tsunami, which struck Japan in 2011, destroying or damaging almost one million buildings, and triggering a meltdown of the Fukushima nuclear powerplant. Cost estimates vary considerably and show a rising tendency as time progresses, but some of the more recent calculations made by independent research institutes suggest that the total costs may be as high as 500 to 700 billion US dollars.[62] However, as a share of GDP, it is not always the case that the most highly developed and highly urbanized societies were hit the hardest: the worst two cases in modern history were the 1988 Spitak Earthquake in Armenia (around 360 percent of nominal GDP) and that of 2010 in Haiti (around 120 percent of nominal GDP), both hitting a relatively small and poor country very hard.[63] Other disasters cost much smaller shares of GDP. This argument can also be extended temporally: while the absolute costs for pre-twentieth-century societies were lower, in relative shares of GDP, many earthquakes were devastating. The 1755 Lisbon Earthquake, with the ensuing tsunami and fire, for example, made two-thirds of the city uninhabitable, destroyed 86 percent of all church buildings, and resulted in the loss of large sums in gold, silver, diamonds, coins, and furnishings. A recent reconstruction suggests that the damage came to between a third and a half of total Portuguese GDP.[64]

---

[58] Soens, 'Resilient Societies,' 154.
[59] Piketty, *Capital*, 106–109, 146–149; Scheidel, *The Great Leveler*, 146–148.
[60] Daniell, Wenzel & Khazai, 'The Cost of Historic Earthquakes Today.'
[61] Alexander, 'The Study of Natural Disasters,' 285. See also Section 2.2.
[62] Behling *et al.*, 'Aftermath of Fukushima,' 414.
[63] Daniell, Wenzel & Khazai, 'The Cost of Historic Earthquakes Today.'
[64] Pereira, 'The Opportunity of a Disaster,' 473–477.

6.1.3 Economic Crisis

A final measure of the short-run material effect of disasters is the economic impact. Disasters can trigger a temporary decline in GDP levels leading to economic crisis: Hurricane Maria reduced Puerto Rico's economy by 3 percent in 2018.[65] The pleas to invest more resources to halt climate change are often based on predictions of dropping economic performance levels and high social and economic costs. Impact on GDP, however, is not always straightforward to measure or analyze. If the destruction is followed by (international) relief, the rebuilding of houses, and the repair and possibly improvement of the damaged infrastructure, the effect may be positive rather than negative. For contemporary disasters, the literature is inconclusive: while some have argued that only disasters followed by political unrest or instability have a negative impact on GDP per capita,[66] others argue that major nature-induced disasters in low- and middle-income countries unleash significant economic setbacks – up to 6.83 percent of per capita GDP for the top 1 percent of disasters in the period 1979–2010.[67]

The nature and intensity of the disaster, but also its preconditions (see Chapter 3) and the immediate reaction to the disaster, all influence the way it may lead to economic crisis. The historical evidence on this is rather meager, not least because historical GDP data are either absent or available only at aggregate national levels, whereas most disasters affect only parts of countries. The 1755 earthquake, fire, and tsunami in Portugal not only destroyed large numbers of buildings and large amounts of capital, but also caused food prices to surge (by 83 percent for wheat and by 171 percent for barley in 1756–57, compared with pre-earthquake average levels) while wages remained stable, except for those of skilled laborers in the building industry, who for some years received an extra premium due to the extraordinary demand for reconstruction work. In the years following the disaster, the already substantial trade deficit of the country, financed by Brazilian gold, widened, as massive amounts of construction materials such as Swedish iron had to be imported.[68] At the same time, however, the economic misery forced the Portuguese to reduce their imports of British textiles and other consumables, which in the longer term may have helped, together with the institutional reforms of Pombal, to gain economic independence from Britain and to promote renewed economic growth.[69]

---

[65] Source: World Bank data, https://data.worldbank.org/country/puerto-rico?view=chart.
[66] Cavallo et al., 'Catastrophic Natural Disasters.'
[67] Felbermayr & Gröschl, 'Naturally Negative,' 104.
[68] Pereira, 'The Opportunity of a Disaster,' 478–481.
[69] Pereira, 'The Opportunity of a Disaster,' 491–495. See also Section 6.3.3.

While the 1755 earthquake triggered economic disruption through its destruction of capital (including human capital),[70] other disasters provoked economic crisis because they affected specific economic sectors, especially in so-called 'mono-product' economies highly dependent on a limited range of activities. The 1279–80 sheep scab epizootic in England illustrates this. Sheep scab is an acute and contagious form of dermatitis caused by sheep mites and it often occurs during damp and cold winters. At that time in England, sheep were kept in folds, and the survival rate of the mites and their offspring was higher because of the colder conditions. Accordingly, in eight months almost 50 percent of the total sheep population in England and Wales were killed by the disease.[71] Even sheep that remained fit to be shorn provided fleeces of far inferior quality and the unit sale prices dropped significantly. This had economic repercussions, since fine English wool was one of the most sought-after commodities in Northwest Europe: clips of wool could be sold years in advance, despite the risk of murrain or sheep scab. When the disaster struck, many Cistercian abbeys could not provide the wool they were committed to supplying, and were forced into bankruptcy. The epizootic, however, had even more far-reaching economic effects. The 1280s was the single worst decade of the thirteenth century because the implosion of wool production affected wool prices, textile production, and international trade. In the second half of the thirteenth century, GDP per head in England is estimated to have shrunk from $828 to $679 because of the sheep scab episode. Between 1277 and 1285 national income contracted by a quarter, showing how crucial the production of wool was for the value of the national economic output.[72] After 1285 wool production and the English economy as a whole recovered.[73]

The economic crises resulting respectively from the 1755 Lisbon Earthquake and the 1279–80 sheep scab epizootic were followed by renewed economic growth in subsequent years. But sometimes the impact of a disaster was more lasting, resulting in a full-blown economic depression lasting for decades or even a century. On a regional scale, we could point to the impact of the Dust Bowl in the United States in the 1930s. Because of the loss of up to 70 percent of the productive topsoil, the affected states that were predominantly agricultural producers took a big hit. Even within the time frame of the Great Depression, the Dust Bowl seriously affected the economy, and population and land values took

---

[70] Argument for effects on human capital also made for the 1629–30 North/Central Italian plague: Alfani & Murphy, 'Plague and Lethal Epidemics,' 335.
[71] Slavin, 'Epizootic Landscapes.'
[72] Campbell, *The Great Transition*, 167. The dollar values are 1990 dollars, commonly used in GDP per capita reconstructions.
[73] Broadberry et al., *British Economic Growth*, 206–207.

decades to recover (see Section 6.1.2). Some scholars have even argued that terrible disasters in the past have led to lasting economic divergences between regions. For example, the severity and spread of the 1629–30 and 1656–58 plagues in Italy caused major economic setbacks through declines in internal demand and destruction of human capital, and according to Guido Alfani may have been responsible for the shift in economic center of gravity away from the Mediterranean to Northwest Europe.[74]

Of course, explaining economic crisis as well as growth requires us to take into account a whole range of variables, and the examples above merely illustrate that nature-induced disasters can play a role in this process – widening an economic crisis which was already unfolding, temporarily damaging vital industries, and paving the way for competitors to enter the market, or triggering institutional responses which in turn have either a beneficial or a damaging impact on economic growth. In history as well as today, the link between disaster and economic crisis tends to be complex and multi-directional.

6.1.4 Scapegoating, Blame, and Social Unrest

Of all types of hazard and shock, epidemics have had the longest tradition of being associated with blame, scapegoating, and even extreme social responses such as violence.[75] Scholars have claimed that "blaming has always been a means to make mysterious and devastating diseases comprehensible and therefore possibly controllable,"[76] and an 'inevitable' component of a pre-modern *mentalité* with poorly formed understanding of causes and cures. This narrative was well entrenched in the 1980s, at a time when societies themselves became preoccupied with HIV/AIDS, leading in turn to widespread concern about the activities of prostitutes, homosexuals, drug dealers, and a general economic 'underclass.' In the 1990s this was further fueled by the 'cultural turn' within the field of history, where fascination with the persecution and burning of witches, for example, sparked interest in researching the role of 'plague poisoners' or 'syphilis spreaders.'[77]

A persistent feature associated with the above-mentioned view is the focus on the capacity of epidemic diseases to loosen the bonds of society,

---

[74] Alfani, 'Plague in Seventeenth-Century Europe.' Similar shocks to demand and long-distance trade via epidemics and famines have been argued for fifteenth-century Europe, ushering in a new age of insular economic practices: Campbell, *The Great Transition*, 17–19.
[75] See the early literature of Baehrel, 'La haine de classe'; Baehrel, 'Epidémie et terreur.'
[76] Nelkin & Gilman, 'Placing Blame,' 362.
[77] Naphy, *Plagues, Poisons, and Potions*; Sidky, *Witchcraft*, 90–91; Bever, 'Witchcraft,' 573; Ross, 'Syphilis.'

disrupt communities from their 'normal' patterns of life, and create unrest.[78] For the Black Death of 1347–52, scholars have focused on the amplification of social tensions, where lawlessness, thefts, and violence were on the up.[79] This is often presented as a 'Boccaccian' breakdown in morals and values – priests, medics, and law-enforcement officials refusing to interact with the afflicted, family members abandoned to their fates, and unscrupulous types waiting around like vultures to appropriate the goods of those who have passed away.[80] Accordingly, it is also unsurprising that a 'Foucauldian' narrative became dominant – epidemics were seen as a 'tool' for powerful elites and authorities to repress and persecute weaker or disenfranchised members of society.[81] Urban governments often used the plague as an opportunity to crack down on those viewed with suspicion, such as vagrants, beggars, and prostitutes,[82] while church authorities used epidemics – but also floods and harvest failures – as a sign of divine retribution for ostentation, greed, and display, legitimizing calls for frugality, piety, or general adaptations of societal behavior.[83] Some of these aspects of disease psychology are still discernible today, with fears perhaps in some regards even heightened through globalization and the notion that 'exotic' diseases can be transported into 'modern' urban environments.[84] In recent times we have seen Muslims blamed for poisoning water systems during the 1994 Surat plague in India,[85] and witnessed the victimization of Asian populations in Chinatowns of various Western cities in the wake of SARS.[86]

Aside from epidemics, a cultural turn within the general field of disaster history also led to increased interest in scapegoating and blame. A notorious example is the wave of witch hunts which appear to coincide with some of the coldest decades of the Little Ice Age: between 1570 and 1650 a remarkable peak in witch trials and convictions occurred in large parts of Europe.[87] While most cultural historians have pointed to juridical, gender, political, and confessional reasons for this period of mass convictions, Behringer and Pfister proposed a causal link between extreme weather – leading to repeated harvest failures and floods – and witch hunts.[88] In this period of climatic extremes, cool and wet summers

---

[78] Delumeau, *La peur en Occident*, 145–165.
[79] Shirk, 'Violence and the Plague'; Bowsky, 'The Impact of the Black Death'; and for other plagues: Pastore, *Crimine e giustizia*; Rose, 'Plague and Violence.'
[80] Biraben, *Les hommes et la peste*, I, 117.   [81] Foucault, *Discipline and Punish*, 195–200.
[82] Lis & Soly, *Poverty and Capitalism*, 79.
[83] Schama, *The Embarrassment of Riches*, 46–48; Akasoy, 'Islamic Attitudes.'
[84] Covello, von Winterfeldt & Slovic, 'Risk Communication.'
[85] Barrett, 'The 1994 Plague.'   [86] Eichelberger, 'SARS and New York's Chinatown.'
[87] Briggs, "Many Reasons Why."
[88] Behringer, 'Climatic Change and Witch-Hunting'; Pfister, 'Climatic Extremes.'

were commonplace, sometimes with snowstorms, floods caused by glacial melt, and hailstorms destroying entire harvests. Although Le Roy Ladurie claimed that climatic changes had only minor effects on society, Pfister argues that recurring events of this nature have a fundamental impact on the mentality of a society. While urging caution against a deterministic link between the weather events of the Little Ice Age and witch hunts, Pfister makes the point that cross-regional surges in witch hunts cannot be explained simply by confessional issues and juridical laxity in the regions affected.[89] The causal links between climate and witchcraft can be found in court records and ego-documents, where accusations of 'weather magic' were common and included charges such as destroying growing crops, causing harvested crops to rot, and spreading diseases among animals. Before 1570, extreme weather events had urged communities to take juridical action, but most cases were dismissed. After 1589, however, large-scale witch hunts were organized. Local communities – collectives, for example – demanded juridical action against unidentified groups of witches blamed for causing the hazards.[90] In the famine year of 1649–50 in Scotland, large witch trials were held; the Scottish parliament claimed that the sin of witchcraft had increased daily and issued 500 commissions to try suspected witches.[91]

Comparable forms of scapegoating associated with extreme weather events can be found in sub-Saharan Africa, linking up with the practice of rainmaking discussed earlier. In the nineteenth-century Zulu kingdom, for example, severe and protracted drought events were often linked with different forms of social unrest by way of the apportionment of blame by the Zulu leadership.[92] In the early nineteenth century, rainmakers – special doctors who attempted to influence supernatural forces through the manipulation of rain medicines – were called upon to bring rain in times of drought; however, when this failed during a protracted drought in the 1820s, rainmakers across the region were killed and Shaka, the Zulu ruler, appropriated control of rainmaking rituals to strengthen his position as a link to ancestral spirits. This position was nevertheless a perilous one, as the ruler risked rebellion if he did not deliver rain. Thus, when severe, multi-year drought struck the region again in 1861–63, further forms of social unrest become widespread in the documentary and oral record. In particular, accounts are replete with reports of 'outsiders' accused of 'nailing the ground' by driving wooden pegs or metal nails into hilltops, apparently to prevent rain. While the origin of this practice is unclear, what is noticeable is

---

[89] Le Roy Ladurie, *Times of Feast*, 119; Pfister, 'Climatic Extremes.'
[90] Behringer, 'Climatic Change and Witch-Hunting.'
[91] Parker, 'Crisis and Catastrophe.' [92] Klein *et al.*, 'Climate, Conflict and Society.'

that it was used to transfer blame from the leadership to outsiders or minority groups. This gave the Zulu ruler an explanation as to why he no longer controlled the skies to the benefit of his people, while the violent response to these minority groups also generated fear of questioning the authority of the kingdom.

Fear, blame, and scapegoating – perhaps even leading to violence – have, then, been for a long time closely associated with historical disaster outcomes in the short term. However, in recent years, research has also started to slowly revise some of these established views – particularly with regard to epidemics. Some have gone as far as to say that epidemic disease outbreaks could also lead to greater bonds of cohesion and compassion.[93] So, while the Black Death of 1347–52 did indeed lead to the most extreme persecution of Jewish families, who were rounded up and burnt to death in the worst cases,[94] this experience was not necessarily replicated in later recurring plagues of the Second Pandemic. Sixteenth- and seventeenth-century commentators in Europe often displayed pride in the funds they raised in cities to aid the afflicted poor, and festivities and ceremonies developed in some regions as a celebration of solidarity.[95] Even for the Black Death itself, recent literature has shown examples of continued compassion and professionalism during epidemics,[96] and quantifiable evidence disputes that there was an increase in criminal activity or a breakdown in legislative institutions.[97] The same has been said of leprosy: while a long-held view is that this disease brought exceptional stigmatization in the medieval and early-modern periods, recent revision suggests this has been exaggerated and may perhaps derive from nineteenth-century politicians' attempts to justify their own cruel treatment plans.[98]

Furthermore, even when epidemics did draw societies out of their 'normal' patterns, it does not appear inevitable that this manifested itself in targeting of the weak or attempts at social control from above by elites. Some of the literature has instead shown a moderate and negotiated balance between isolation, quarantine, and medical concerns, on the one hand, and the need for maintaining economic and commercial life, and the continuance of ritual engagements, civic freedoms, and customary community ties on the other.[99] When there was a breakdown in social

---

[93] Cohn, *Epidemics*, 68–92.
[94] Cohn, 'The Black Death and the Burning of Jews'; Colet *et al*., 'The Black Death and Its Consequences.'
[95] Cohn, *Epidemics*, 87–92. In Naples in 1658 there was even an insinuation that collective celebratory festivities had perpetuated the disease: Guarino, 'Spanish Celebrations,' 30.
[96] Wray, *Communities and Crisis*, Chapters 3 and 5; Wray, 'Boccaccio and the Doctors.'
[97] Dean, 'Plague and Crime,' 385.   [98] Rawcliffe, *Leprosy*, 43.
[99] Wilson Bowers, *Plague and Public Health*; Murphy, 'Plague Ordinances,' 144.

order, generally it went in the opposite direction to the 'Foucauldian' narrative, as lower orders used the epidemic as an opportunity to offer either passive or active resistance against visible representatives of elites and authorities – often city officials, medics, or 'plague workers.'[100] Even more famously, in the nineteenth and twentieth centuries, cholera became associated with mass uprisings and violence from below – elements of which have been mirrored in the twenty-first century.[101] A paradox in this context was that disruption from below – although itself constituting a breakdown in social order – originated in an intention to preserve tightly held customs and norms, especially with regard to burial practices. Therefore, it is important to note something persistently linking pre-industrial, modern, and contemporary responses: communal suspicions and distrust of the decisions and actions of 'elites', 'authorities', 'experts,' or 'outsiders' are seen even today in the wake of serious epidemics – the case of the 2014 Ebola outbreak in West Africa is an obvious example, leading to violent resistance and outcry at the local level.[102] The need to engage local communities and be respectful of their cultural contexts when developing disaster response strategies (such as quarantine) is something that not only runs very deep across different societies, but also goes far back in time.

6.2 Societal Collapse

As mentioned already in this book,[103] disaster research in recent years has tended to place more emphasis on the resilience of societies, communities, and individuals in overcoming the challenges presented by hazards and shocks. Societal resilience and adaptation – to differing degrees – have become the 'norm'; even if at the same time certain groups of people within those societies were also differentially vulnerable. Indeed, prolonged crises leading to total societal collapses – with systemic dysfunction – have been shown to have been rare in historical perspective. Nevertheless, these extreme occurrences did happen. Although a commonly accepted definition of societal collapse is hard to come by, many scholars agree that it represents a rapid, fundamental transformation of the social, political, and

---

[100] Curtis, 'Preserving the Ordinary.' A broader thesis on plague as a catalyst for popular social unrest is presented in Cohn, *Lust for Liberty*, Chapters 9 and 10.

[101] Snowden, 'Cholera'; Briggs, 'Cholera and Society'; also for Third Pandemic plague outbreaks: Lynteris, 'Suspicious Corpses.' For the recent social unrest in the wake of post-earthquake cholera in Haiti, which was seen as brought to the country by 'outsiders' (the UN): Farmer, *Haiti*, Chapter 7.

[102] Kutalek *et al.*, 'Ebola Interventions'; Pellecchia *et al.*, 'Social Consequences of Ebola'; Calain & Poncin, 'Reaching Out to Ebola Victims.'

[103] See Section 2.3.4.

economic structures of a complex society for multiple generations.[104] Often these transformations have fundamental effects on the environment, on the population, and on ideology, values, and belief systems.

Although we view societal collapse as being historically rare, older historical literature tended to see the process as something that was almost inevitable: all 'great' societies and civilizations rise but eventually fall. This paradigm arguably began with Edward Gibbon and *The History of the Decline and Fall of the Roman Empire* (1776), connecting this fall with the idea of 'moral decay.'[105] According to the narrative, republican spirit, modesty, and militarism gave way to a decadent society focused on luxury and pacifism, creating fundamental vulnerabilities towards invasions and, in turn, collapse. This belief in moral decay as intrinsic to societal collapse had already been proclaimed by Ibn Khaldun, writing in the fourteenth century, when he described the rise and fall of the Islamic Empire. In the nineteenth century, archaeological discoveries of past civilizations that had collapsed served to enhance these views. 'Egypt mania' and explorations of ancient Babylon and Assyria, and the rediscovery of the Mayan and Inca capitals, showed that historical societies that had blossomed for centuries all ultimately reached a peak and then seemingly abruptly collapsed. The cyclical nature also corresponded to Darwinian insights that all organisms and species go through the stages of growth, maturity, and decline.

Recent literature still discusses societal collapses, but they are no longer explained by moral standards – referred to by Tainter as 'empirically unknowable' or 'unobservable' factors – and instead are commonly linked to environmental hazards, including resource depletion as well as climatic changes and tectonic hazards, and the disasters that can ensue.[106] For example, weather extremes or larger-scale climatic shifts have been central in explaining the multiple 'disappearances' of ancient societies. The sudden and simultaneous collapses of the Ancient Egyptian, Indian, and Mesopotamian societies have been linked with climatic anomalies and extreme drought around 4000 years ago,[107] while a similar occurrence of drought has been linked with the Maya Terminal Classic collapse in 900 CE, and the decline of the states centered around Mapungubwe and Great Zimbabwe during the early to middle part of the second millennium CE in Southern Africa.[108] The collapse of the Western Roman

---

[104] Definition based on Butzer & Endfield, 'Critical Perspectives'; Luzzadder-Beach, Beach & Dunning, 'Wetland Fields,' 3646; Tainter, *The Collapse of Complex Societies*.
[105] Gibbon, *The History*. [106] Tainter, *The Collapse of Complex Societies*.
[107] Dalfes, Kukla & Weiss (eds.), *Third Millennium BC Climate Change*.
[108] The classical perspective is described and revised in Luzzadder-Beach, Beach & Dunning, 'Wetland Fields'; Holmgren & Öberg, 'Climate Change,' 185–195; Huffman, 'Climate Change during the Iron Age.'

6.2 Societal Collapse

Empire is nowadays presented as an outcome of climatic instability and the detrimental effects of terrible epidemics such as the Antonine and Cyprianic Plagues.[109] Particular episodes such as the Little Ice Age have also been posited as the explanation for societal collapses – most notably the demise of the Norse society in Greenland, which has been in part attributed to the general cooling and therefore inhospitable conditions of the late fourteenth and fifteenth centuries.[110]

Nevertheless, environmental explanations of societal collapse have attracted some serious critiques. First, many of these studies have been labeled environmentally deterministic, because of the mono-causal nature of explanations and the simplistic image that is often painted – one that is often driven by the signals of paleoclimate data, the dating of which is often uncertain.[111] Increasingly, however, emphasis has been placed on the co-agency between weather extremes and societal transformations rather than on direct causality.[112] Second, the causal relationship between environmental hazards and eventual collapse is not easily established. Several societies in decline, such as the Ancient Egyptian, Mesopotamian, and Indian civilizations, did not actually collapse simultaneously during a single drought event, but declined over the course of more than two centuries – requiring a more nuanced understanding of the temporal dynamics.[113] This point also relates back to issues with the definition of collapse itself. For some scholars, population numbers and societal complexity outweigh forms of cultural survival such as language and religion, while others may hold art styles and literary traditions as counter-evidence for collapse. Thus identifying cases of 'collapse' is as much a value judgement over what one considers as success and failure as an empirical one.[114]

Equally fundamental is the point raised by Tainter – writing three decades ago – that attempts at explaining collapse have often descended into primarily factual contests around whether particular pieces of historical or archaeological evidence either support or contradict a certain position. In his view, and this is arguably still the case in some of the literature on this topic, this has been at the expense of more careful consideration of the logic of the original proposition around collapse. This involves more basic questions such as "how can environmental or

---

[109] Harper, *The Fate of Rome*.
[110] The classical perspective is described and revised in Dugmore *et al.*, 'Cultural Adaptation.'
[111] Middleton, 'Nothing Lasts Forever.'
[112] Warde, 'Global Crisis'; Degroot, *The Frigid Golden Age*.
[113] Luzzadder-Beach, Beach & Dunning, 'Wetland Fields.'
[114] See arguments in McNeill, 'Sustainable Survival.'

Figure 6.2 Painting by Thomas Cole, *The Course of Empire – Destruction* (1833–36). Gift of The New York Gallery of the Fine Arts.

climatic factors lead to collapse?" and "can these variables really account for the eventual outcome?" If these questions cannot be answered convincingly, then any discussion of evidence becomes a distraction.

Evidence also suggests that decline usually does not result in complete collapse, but rather in the eventual transformation into a fundamentally new type of society. Total abandonment such as that experienced by the Norse society of Greenland is extremely rare and represents an anomaly in the course of history. Although climate and disease have been linked with the decline of the Western Roman Empire by some authors, others have tried to point to elements of adaptation, transition, and continuity instead.[115] In the case of Ancient Egypt, the two intermediate periods were not sudden collapses between periods of grandeur, but were preceded by significant societal transformations, political problems, and social unrest that eventually led to a period of turmoil and a new societal state. This can hardly be called a societal collapse, since some social groups or parts of society showed signs of resilience and adaptation that are masked by the often grand-scale political changes.[116] Overall, then, we should make it clear that societal collapse was the exception rather

[115] Wickham, *The Inheritance of Rome*; Haldon *et al.*, 'Plagues.'
[116] Butzer, 'Collapse, Environment, and Society.'

than the rule throughout history – and even some of the so-called 'classic' collapses may be conceived of more as transitions and adaptations rather than as the destruction of all social, economic, and political structures. That is not to say, however, that we should downplay the severity and trauma of these reconfigurations – indeed, even though we regard them as transformations rather than collapses, these processes still went hand in hand with large social costs, especially for the most vulnerable segments of those societies.

6.3 Long-Term Effects

Until now we have zoomed in on mitigation strategies and short-term effects and recovery, but as historians we also need to draw attention to long-term effects that are either frequently overlooked or impossible to foresee or measure for very recent disasters. The immediate link is not always that clear, and other societal factors can interfere as well. Nevertheless, these slower processes and long-term effects are critical, especially in cases of recurrent or repetitive shocks. In general three outcomes are possible: recovery, stagnation, or decline. Currently much of the focus lies on societies that adapt after a hazard or shock and therefore are able to recover in the long run, but alternative scenarios are possible too. These different paths are discussed over the course of this section.

6.3.1 Disasters as a Force for Good? Economic Effects

In some cases, disasters have been seen as 'positive shocks' that stimulate economic changes. Although, as iterated above, adaptation was never inevitable in the aftermath of a disaster, sometimes these events did become a force for good. Earthquakes that destroyed large parts of cities caused significant damage to people and capital, but other, 'positive' consequences could be the complete re-planning of the city, making it safer and healthier,[117] or the increased demand for employment in the building and laboring industries – also a feature of flooding. Some scholars have pointed to new property reforms, adaptation of antiquated inheritance laws, and modernization of land markets as some of the economic 'benefits' to come out of the 1755 Lisbon earthquake.[118] As mentioned already, people could learn from disasters, developing new

---

[117] Condorelli, "*U tirrimotu ranni*," 331–353; Andreau, 'Histoire des séismes.'
[118] Pereira, 'The Opportunity of a Disaster,' 467; with a response in Aguirre, 'Better Disaster Statistics'; and some disagreement in Serrão & Santos, 'Land Policies.'

policies, institutions, and infrastructures that not only protected societies more securely from future hazards or shocks, but also entailed more widespread gains for overall welfare and development.

What we need to ask, however, is whether short-term reconstruction always equates to long-term economic recovery. Evidence tends to suggest that long-term aggregate economic developments after disasters were quite diverse. In his latest book *The Great Transition*, Bruce Campbell elaborates on the distinct paths that Western Europe and Asia took from the second half of the thirteenth century onwards, as the then known world came to be hit by a series of epizootics, harvest failures, epidemics, and weather extremes within the broad framework of cooler and unstable global climatic conditions. This occurred simultaneously – perhaps causally – with a series of revolts, wars, and mass migrations, one of the most famous being the Mongol invasion of Genghis Khan, which affected different empires and societies all along the network of the Silk Road. While many parts of China were far superior to Europe in technological and economic development before this Great Transition, the aftermath saw China's star wane through repeated wars and environmental distress. According to Campbell, this outcome could not possibly be explained by the economic, political, or social trends that manifested themselves at the start of the fourteenth century. Only by looking at the combined effects of war, disease, and environmental change and the coordinated responses that were taken by the different societies can this divergence be explained. The calamitous fourteenth century, therefore, was pivotal in shaping the long-term economic divergences of East and West.[119]

With specific regard to the long-term economic impact of epidemic diseases, there has been a lengthy debate – still going on today – as to whether they produced 'positive' or 'negative' development outcomes. For example, a large body of literature has tended to suggest that the Black Death of 1347–52 had positive economic effects, mainly through channels of redistribution described below: a destruction of labor, but keeping capital intact, and therefore improving the lives of the survivors who saw real wages rise and elements of extra-economic coercion subside.[120] Indeed, some scholars have suggested that the places where the mortality effect of the Black Death was harshest eventually experienced the most favorable long-term economic trajectories.[121] Yet we also have to ask ourselves to what extent these 'favorable' redistributive outcomes were negated by elements of aggregate contraction: smaller

---

[119] Campbell, *The Great Transition*, 19–30.   [120] Pamuk, 'The Black Death.'
[121] Voigtländer & Voth, 'The Three Horsemen.'

economies with fewer vacancies and more insular and contracted patterns of trade.[122]

The problem with coming to a coherent answer on the long-term economic effects of epidemics is that, quite simply, these diseases often meant different things to different people. For many city dwellers in Western Europe after the Black Death, credit became easier to obtain and interest rates dropped, yet for many rural people life became more expensive, with only a few exceptions.[123] Real wages may have increased after the Black Death in many places, but was it the case that everyone benefited from this? Recent research suggests that women did not do so to the same degree as men.[124] A further potential limitation is that the economic effects of historical epidemic diseases have tended to be very rigidly measured with 'traditional' economic indicators such as GDP, urbanization, wages, property distribution, fertility, and population recovery. Yet many of these indicators are still based on low-resolution macro-level data, sometimes with large chronological gaps between epidemics and economic/demographic markers, and often available only for restricted segments of society – for example, adult males of a certain skill level in an urban environment. The economics and development studies literature also uses these same indicators for contemporary societies, but integrated within a much broader set of economic markers after epidemics that include care, protection, knowledge and skills, health, social networks, and isolation. This is a recognition that epidemics did not just kill people, but killed people of different statuses and skills, and they offered new opportunities and put new pressures on those that survived.[125] Some of the most vulnerable became more exposed, some workers and care-givers became more burdened, and new responsibilities were sometimes thrust upon those lacking skills and experience. Accordingly, it is not surprising that, while the historical literature still tends to frame disease–economy interactions in a positive light,[126] almost all of the literature in economics and development studies suggests that disease outbreaks affect economies negatively – particularly with regard to the impact of malaria and HIV/AIDS on productivity and human capital formation.[127]

---

[122] Alfani & Murphy, 'Plague and Lethal Epidemics'; Campbell, *The Great Transition*, 355–373; Álvarez-Nogal & Prados de la Escosura, 'The Rise and Fall of Spain.'
[123] Van Bavel, *The Invisible Hand?*, 109.
[124] Humphries & Weisdorf, 'The Wages of Women.'   [125] Madhav et al., 'Pandemics.'
[126] Acemoglu, Robinson & Johnson, 'Disease and Development.'
[127] Bloom, Canning & Fink, 'Disease and Development Revisited'; Sachs & Malaney, 'The Economic and Social Burden of Malaria'; Bell & Lewis, 'Economic Implications of Epidemics.'

Nevertheless, for many other types of disasters, even if the long-term economic impact was negative, the scale was generally local or at best regional. The farm land lost when in the tenth century the village of Kootwijk (the Netherlands) was buried by sand was never recovered. In fact, in the twelfth century changed climatic conditions led to the enlargement of the 'sand desert' that now came to encompass a number of pre-existing drift sand nuclei. Still, the economic effects remained restricted to the area directly affected by the drift sand.[128] Moreover, there is often a lack of clarity over exact causal links between disaster and economic outcome. One of the most significant economic downturns was connected to the collapse of the water management systems around the Nile in Mamluk Egypt, though at the same time, even the roots of this breakdown were connected in the first place to epidemic mortality.[129] Similarly, significant floods of the Yellow, Yangtze, and Huai Rivers during the nineteenth and early twentieth centuries have been established as integral facets to the narrative of environmental pressures and Chinese economic difficulties after 1800,[130] though again it is difficult to separate the economic effects of floods from the famine conditions that also often followed.[131]

6.3.2 Long-Term Demographic Changes

As well as varying in terms of their economic recoveries, disasters also showed very different degrees of population recovery. We have seen that the mortality effects of famines could diverge sharply, not only from region to region, but also between localities within regions – according to whether or not famine-related diseases such as dysentery, tuberculosis, or typhus emerged. In fact, long-term recovery after famines was dictated by trends influencing nuptiality and fertility.[132] In pre-modern times, during the course of the famine people actually tended to delay their marriage, and fertility conditions were understandably sub-optimal. In the aftermath of famines, however, just like with other mortality shocks, there were often spikes in marriages – or rather remarriages. This was a tendency – not a hard-and-fast rule – and so it was neither inevitable nor predictable across pre-modern societies. In some places, remarriage after widowhood was culturally restricted – especially for women, as in the Kingdom of Naples, for example.[133] The rates of remarriage also depended on pre-existing

---

[128] Heidinga, 'The Birth of a Desert'; see also Section 6.1.2.
[129] Borsch, 'Plague Depopulation'; Borsch, 'Environment and Population.'
[130] Elvin, *The Retreat of the Elephants.*   [131] Li, 'Life and Death.'
[132] Galloway, 'Basic Patterns.'
[133] Curtis, 'An Agro-town Bias?'; see also Section 6.1.1.

6.3 Long-Term Effects

institutional configurations – inheritance practices, dowry demands, and access to and control of property.[134] Many families did not want to risk fragmentation of their estates, and thus newly widowed adults, often women, were urged to remain single. Arguments have also been made around the importance of 'cultural flexibility' for recovery from demographic disaster – with particular reference to indigenous societies in the Americas following substantial population losses during the fifteenth and sixteenth centuries. The Tapirapé in Brazil, for example, experienced protracted population decline following European contact, partly as they continued to observe marriage rules that acted to control population growth, whereas the more flexible marriage rules of the Tenetehara, which promoted population increase, enabled recovery.[135] Accordingly, these differential pressures on the rates of nuptiality after famines and epidemics led to differential fertility responses too. This was particularly the case when a large degree of famine migration had taken place, by those in search of resources, food, and work, and especially if this was sex-selective migration. In some rural areas men left women and children behind, and did not return – creating sex-skewed habitation in their places of origin and impacting upon marriage opportunities.

One of the most significant debates on the population effects of disasters remains connected to the Black Death and recurring plagues. Of course, there is no debate that the Black Death killed many people – numbers that have been revised upward in recent years.[136] However, there were great discrepancies between different regions as to the demographic recovery rates. It is widely accepted in the literature that, in this context, the Low Countries (for example) recovered very quickly and some areas there had already reached pre-Black Death levels at some point in the fifteenth century. Iberia and Italy did not achieve this position until the sixteenth century, and in England the balancing point came as late as the seventeenth century.[137] A further complication is that recent literature has also questioned some of the empirical evidence for these population recoveries – for example, in the case of the Low Countries, assertions have been made that are based on a paucity of quantifiable information, and often rely on sources that cannot distinguish between population trends and rural–urban migration trends.[138]

---

[134] Guinnane & Ogilvie, 'A Two-Tiered Demographic System.'
[135] Newson, 'The Demographic Collapse.'
[136] Benedictow, *The Black Death*, 383; Alfani & Murphy, 'Plague and Lethal Epidemics'; Lewis, 'Disaster Recovery'; Campbell, *The Great Transition*, 306–319.
[137] Van Bavel & van Zanden, 'The Jump-Start'; Malanima, 'The Economic Consequences'; Broadberry et al., *British Economic Growth*, 20, 29.
[138] Roosen & Curtis, 'The "Light Touch."'

Setting aside concerns over the empirics, however, scholars have debated the actual causes of the differential rates of recovery. Some have emphasized mortality: that is to say, the late-medieval demographic context was a high-mortality regime dictated by severe repeat epidemics, and this contrasted with a more favorable lighter-mortality regime in the early-modern period.[139] Unfortunately, there are three problems with this scenario. First, our empirical evidence to compare mortality regimes between late-medieval and early-modern periods is scarce. Second, it goes against other literature emphasizing the insalubrious nature of the early-modern cities as 'urban graveyards.' Finally, it does not explain geographical divergences in recovery rates, unless we suggest that mortality regimes were differential across areas – something difficult to do given the point made above about lack of quantifiable evidence. It has been suggested that the major change was, in actual fact, mortality driven by urban-based diseases, in contrast to the Middle Ages, when epidemics spread more widely: however, again, there is little systematic evidence for this, and recent work has shown territorially pervasive epidemic outbreaks in both late-medieval and early-modern environments.[140]

6.3.3 *Reconstruction, Reform, and Societal Change*

As we have seen, the effects of a hazard or shock can differ significantly, depending on the societal response both in the immediate aftermath as well as in the long run. One of the options is not to act at all, because of either the impossibility of responding or an unwillingness to respond. Another option is a societal adaptation that is ill-suited to the type of hazard or to the society itself, which might lead to a societal collapse or prolonged crisis. In this section, however, we focus on types of long-term effects and strategies that often follow after a large-scale disaster: reconstruction and reform, and pushing for societal change.

Reconstruction and Reform One of the ways in which disasters lead to long-term changes is via their capacity to stimulate new forms of reconstruction and reform – either of physical environments or of societal institutions. For example, societies responded in a very structural way to outbreaks of diseases, and this changed over time. Recent literature has shown how medieval societies in Europe were already implementing forms of environmental control in the fourteenth and fifteenth centuries –

---

[139] Hatcher, 'Mortality'; Hatcher, *Plague*, 64; Benedictow, 'New Perspectives'; Cipolla, *The Economic History*, 77; Flinn, 'The Stabilisation,' 286.

[140] Roosen & Curtis, 'The "Light Touch"'; Curtis, 'Was Plague an Exclusively Urban Phenomenon?'

6.3 Long-Term Effects

especially regulating and sanctioning practices seen to be unhygienic or in contravention of what was seen to be in the interest of public health, or to prevent environmental pollution.[141] Still, of course, much of this was based around societies' view that diseases spread through 'miasma' and were linked to bad 'auras,' dampness, or smells. When these environmental controls combined with other public health interventions such as effective quarantine systems, the number of future epidemics could be reduced: Ragusa (Dubrovnik), for example, experienced its very last 'domestic' plague in 1533,[142] comparing favorably with parts of Western Europe which continued to experience Second Pandemic plagues into the seventeenth century, and other parts of Southeast Europe which were not plague-free until the eighteenth century.

After the demographic or epidemiologic transition, mortality rates lowered and life expectancies increased in certain parts of Eurasia and North America – partially testament to the waning or even disappearance of Second Pandemic plagues – but in the nineteenth century, infectious and contagious diseases were still present, especially cholera, smallpox, influenza, and tuberculosis, and often linked to over-populated, industrializing urban environments. As well as instigating short-term mitigation measures such as sanitary cordons and hospitalization – clear links to quarantines of the past – these urban-focused diseases also helped stimulate more fundamental long-term strategies. This went hand-in-hand with developments in scientific and medical knowledge – the so-called 'laboratory revolution' – as miasma theories of contagion were replaced by a better understanding of the role of bacteria and contamination in what became known as 'germ theory.' So, for example, by the 1860s, the link between cholera outbreaks and contaminated water had given rise to sanitary reforms, where water pipes and sewers were replaced.[143] Antiquated cesspool and privy-vault systems for waste removal were replaced by underground gravity-flow systems – necessitating large-scale capital investments at the same time.[144]

Reconstruction and reform policies were, and still are, also a common long-term response to other types of disasters than epidemics. In earlier chapters several examples have come up, such as the introduction of agricultural innovations, early warning systems and food-for-work programs in Ethiopia and Sudan after the famines of the 1970s and 1980s,[145] or the development in the North Sea region of economic strategies that

---

[141] Rawcliffe, *Urban Bodies*, 12–15; Coomans, *In Pursuit of a Healthy City*, 91.
[142] Blažina Tomić & Blažina, *Expelling the Plague*, 62–63.
[143] Baldwin, *Contagion and the State*, 147–156.
[144] Tarr, *The Search for the Ultimate Sink*, 112–117.   [145] See Section 5.2.1.

allowed communities to live in a tidal landscape with frequent inundations.[146]

Pushing for Societal Change Disasters both past and present have been used to make the case for social change – even in some cases revolution. The communist revolutions in Russia (1917) and China (1949) took place during or after periods of serious food crises, which in turn can be connected to the earlier establishment and subsequent growth of new capitals – St. Petersburg and Beijing – in strategically important but food-deficient regions. Feeding those capitals and the rapidly growing armies amassed by both empires became increasingly difficult. In Russia, it required regular transports of very large quantities of wheat from the Black Earth and Volga regions to the North, first by river and later by train. During World War I the combination of wartime requisitions, the German occupation of Ukraine, and transport blockades gave rise to acute food shortages in the North. In China the crisis was drawn out over a longer period of time. The provisioning of Beijing relied on state-organized transports of tribute rice from the South to the North. This state grain supply system increasingly came under pressure from the second half of the nineteenth century as a result of rebellions, wars with the British, silting of the main waterways, and rapid population growth. The severe famines which ensued in the North fed discontent and contributed to the struggles that ultimately led to the communist take-over. Both in Russia and in China the newly established communist regimes considered an escape from the food supply problems in the preceding 'times of troubles' as one of their main challenges.[147]

In *The Shock Doctrine*, Naomi Klein argues that modern Western societies, and especially capitalistic ones, have used disasters of all kinds, such as war, terror attacks, hurricanes, or tsunamis, to push through fundamental societal changes that could not happen in a 'normal' or non-crisis situation. She attributes this way of thinking to one school of thought and especially to one man, Milton Friedman, spokesman of the Chicago School of economics, who stated that "Only a crisis – actual or perceived – produces real change." According to this theory, the sudden, unexpected nature of a crisis creates a numbness, anxiety, and compliance within the general public that are ideal conditions for the authorities to implement policies that fit their agenda. A case in point is the development of charter schools in New Orleans after Hurricane Katrina in 2005. Friedman used the opportunity of the destruction of most New Orleans schools not only to rebuild the physical infrastructure, but also to change the education

---

[146] See Section 4.2.1.   [147] Wheatcroft, 'Societal Responses.'

system at the same time. Referring to a 'clean slate,' Friedman proposed changes from a public-school-dominated policy towards charter schools and individual household education vouchers that could be spent in privately owned and managed schools.[148]

While Klein focuses on the contemporary period and neoliberal states, this principle has much deeper historical precedents. After the 1755 Lisbon Earthquake, the Marquis de Pombal took the opportunity to revolutionize the Portuguese economy and policies. Besides clearing the rubble and reconstructing buildings, de Pombal also rearranged property structures, reformed the tax system, and attacked the power of the Church and nobility, in order to modernize the feudal and rural state. It was the shock of the unforeseen earthquake that provided the unlimited power and the ability to break through former resistance both from the nobility and from the rural population.[149] Going further back in time to the early-modern period, it has been suggested that epidemics provided the ideal social circumstances for authorities not only to impose restrictions such as quarantines – but to go much further and extend their influence into the lives of ordinary citizens, urging, for example, other forms of social control around gatherings, meetings, and public spaces, and even criminalizing certain forms of behavior.[150]

These shock-induced societal shifts were not uniformly so well planned and organized, however. After the calamitous fourteenth century in Northwest Europe the mortality crisis made labor-extensive policies more favorable than labor-intensive grain production. As a result, in large parts of Europe manorial lords and yeoman farmers switched over to cattle and sheep breeding on extensive pastures, thereby fundamentally altering the medieval economy. Common land was enclosed or redefined, rent systems were altered, and labor conditions were revised. In this case, there was no single identifiable figure – no Friedman or de Pombal – leading the policy shift. The long-term effects and changes were implemented more gradually and were created because of a shift in relative prices.

Shocks did not always lead to significant societal change, however: even in the case of major disasters, resistance might well be too strong to be overcome. In the years before the disaster at the nuclear plant of Fukushima, little weight had been given to the risks connected to the possible occurrence of a tsunami. A number of factors contributed to this culture in which risks were downplayed. They included a perceived lack

---

[148] Klein, *The Shock Doctrine*, 3–24.
[149] Araújo, 'The Lisbon Earthquake'; Pereira, 'The Opportunity of a Disaster.'
[150] Curtis, 'Preserving the Ordinary.'

of alternatives to nuclear energy, the desire to preserve public acceptance by presenting nuclear energy as inherently safe, and close relations between officials at the Nuclear and Industry Safety Authority (NISA) and the industry – including the *amakudari* system (the custom whereby retired NISA officials took up advisory positions in industries they had supervised during their careers). Research only a few years before the disaster took place had pointed out the need for a higher seawall in this earthquake-prone region – yet was ignored.[151] The Fukushima disaster sent shock waves through Japanese society, giving rise to a call for strict safety regulation of the nuclear industry and, in fact, for a complete overhaul of energy policy. But, although new safety legislation has been passed and NISA has made way for the new, independently operating Nuclear Regulation Authority, drastic strategy reforms did not take place. Almost all nuclear plants were temporarily shut down shortly after the disaster, but pressure to restart some of them was, and still is, strong. The 'Innovative Strategy for Energy and Environment' formulated shortly after the disaster, aimed at substantially reducing dependency on nuclear energy, met with fierce opposition from businesses and the nuclear industry: the high costs, it was argued, would erode Japan's competitive position on international markets and its eminence in nuclear technology. New plans formulated afterwards were much more ambivalent about the future of nuclear energy, and Japanese society is still deeply divided over the issue.[152]

6.3.4 *Economic Redistribution*

Hazards and shocks can lead to redistribution of economic resources. This view has recently been supported in grand theses by World Bank economist Branko Milanovic, who suggests that "epidemics and war alone can explain most of the swings in [pre-modern] inequality."[153] Elsewhere, in a best-seller in its field, Walter Scheidel has argued that throughout history socio-economic inequalities have leveled themselves out only during episodes of either mass mortality or intense violence and accompanying mass mobilization.[154] These views are important because they downplay the egalitarian effects of active societal intervention and progressive welfare. Reducing inequality by peaceful means appears harder than ever,[155] and is left to the vagaries of sudden events often outside our control.

---

[151] Kingston, 'Mismanaging Risk.'   [152] Duffield, 'Japanese Energy Policy.'
[153] Milanovic, 'Income Inequality,' 480.   [154] Scheidel, *The Great Leveler*.
[155] Scheffer *et al.*, 'Inequality.'

6.3 Long-Term Effects

While major wars are said to reduce inequality through the destruction of capital or the implementation of welfare policies, the logic behind a more 'equitable' redistribution for epidemics tends to be the obliteration of people while keeping capital intact, thereby realigning the economic balance back in favor of labor.[156] So, according to this line of reasoning, in the aftermath of the Black Death, the gap between 'elites' such as lords and aristocrats and the 'lower orders' such as peasants and laborers was narrowed through higher wages, easier mobility, reduced extra-economic impositions, and greater opportunity to purchase property. Another element of this logic, with specific regard to property redistribution after epidemics, is that sudden mass mortality could have some equitable effects, at least in the short or medium term, because in certain conditions of partible inheritance, property was more likely to be divided up and fragmented between different heirs. Furthermore, high epidemic mortality led to cases where elderly adults had nobody to pass their property on to, creating families unable to maintain or consolidate estates for more than one generation.[157] In turn, many post-Black Death societies became less 'unequal' with improved living standards for 'ordinary' survivors.[158]

Although these hypotheses are highly stimulating, we should still approach them with a level of caution. First, it might be said that, although the 'leveling' thesis connecting shocks to capital and labor has popular fascination, it also diverts our attention from general long-term inequality trends. That is, despite the almost constant sequence of epidemics and wars across history, the general rule that wealth is almost always accumulating in the hands of elites has become more and more confirmed. Recent research into pre-industrial inequality levels in the Low Countries, Italy, and Spain has demonstrated that, regardless of the short-term redistributive effects of disasters, the long-term trend of inequality from the fifteenth century onwards was rising.[159] Even the Black Death, proclaimed as one of the most redistributive shocks, only led to heightened equality for less than a century, which was quickly reversed thereafter,[160] and many other severe epidemics had egalitarian effects only for a number of years rather than decades.

---

[156] Milanovic, 'Income Inequality'; Scheidel, *The Great Leveler*, 304–305; Pamuk, 'The Black Death.'
[157] Razi, 'The Myth,' 30.
[158] Dyer, *An Age of Transition?*, 128–139; Cohn, 'Rich and Poor.' Bioarcheological evidence suggests people were healthier after the Black Death: DeWitte, 'Health in Post-Black Death London.'
[159] Alfani & Ryckbosch, 'Was There a "Little Convergence" in Inequality?'; Furió, 'Inequality and Economic Development.' Such a continuous rise in inequality was not matched in Portugal: Reis, 'Deviant Behaviour?'
[160] Alfani & Ammannati, 'Long-Term Trends in Economic Inequality.'

Second, it might also be said that much of this focus on the redistribution of wealth, property, and income after epidemics and wars has tended to obscure other elements of (in)equality. For example, the old view that the Black Death led to a 'golden age for women' in Northwest Europe has been contradicted by new evidence: whatever long-term benefits occurred in the form of rising real wages for men after the Black Death, women did not share in these benefits.[161] Bioarchaeological evidence, furthermore, supports the contention that women did not share in post-Black Death health benefits in the same way as men.[162] Put simply, then, to what extent can we describe a redistributive effect of mortality shocks as 'equitable,' when half the population never shares in the benefits of such redistribution? This argument can be extended to other dimensions outside of gender: for example, regardless of the effects of the Black Death and recurring epidemics in the Kingdom of Valencia, it is clear that Muslims could not scale the feudal hierarchy of the Christian Kingdom, despite comprising a third of the population and exhibiting their own internal social stratification.[163]

Third, there are some issues with the causal mechanisms often invoked. The Black Death is a well-cited example, but also potentially an anomaly in its redistributive effects, or at the very least an extreme case. We should not just assume that all socio-economic responses to shocks mirrored those of the Black Death in a universally applicable model.[164] There was a broad spectrum behind the ratio of capital to labor damage, and this diverged from historical case to historical case. In fact, rather than an 'inevitable' form of redistribution after hazards and shocks, what we tend to see in the pre-industrial period are redistributive outcomes that are not uni-linear, vary in intensity, and do not always last the same amount of time – that is to say, are not always structural changes.

While shocks did, of course, produce some equitable outcomes, this was not always the case – sometimes certain groups were better able than others to buffer these events,[165] testament to their pre-existing resource and power advantages, instrumentalizing the shock to their benefit and in

---

[161] For an outline of the debate on the 'golden age of women' see Rigby, 'Gendering the Black Death,' 745–754; with updated historiographical references also in Kowaleski, 'Gendering Demographic Change.' New male–female disaggregated wage data can be found in Humphries & Weisdorf, 'The Wages of Women.'
[162] DeWitte, 'Stress, Sex, and Plague'; Lewis, 'Work and the Adolescent.'
[163] Baydal Sala & Esquilache Martí, 'Exploitation and Differentiation,' 61, 64.
[164] Alfani & Murphy, 'Plague and Lethal Epidemics.'
[165] For floods: Curtis, 'Danger and Displacement'; For wars and famines: Alfani, *Calamities and the Economy*, 76; Galloway, 'Basic Patterns,' 288; Campbell, 'The Agrarian Problem,' 43.

the process exacerbating inequalities.[166] For example, recent research has shown that an epidemic in 1570s Mexico, which reduced the indigenous population by 70–90 percent, facilitated a process of elite concentration and increased the amount of land farmed through large estates. Where collapse was less severe, indigenous villages were better able to maintain control of communal lands, hindering colonist encroachments.[167] This is yet another important point to make, because the wealth of pre-industrial communities was not always tied up in what the individual or the household owned, but could also be located in collective or common pool institutions.[168]

The role of taxation is a case in point, and may reveal some key differences between the redistributive effects of modern disasters and those of the pre-industrial period. Discussing the effects of the two World Wars, Piketty has pointed out the significance of the policy shifts immediately afterwards. Thanks to progressive taxation, the accumulated wealth of the elites, that had been steadily growing until the nineteenth century, was redistributed fundamentally during the 1940s and 1950s. At the same time, the poorer classes could climb the social ladder because of rising wages, public services, and new social security systems.[169] This redistributive effect of taxes is, however, a contemporary phenomenon. In pre-industrial societies, taxes were not designed to finance public services and redistribute wealth to provide for the poorer social classes in society.[170] A continuous and recurrent tax regime developed only from the sixteenth century in most centralized states of Europe, and prior to this, governments could levy a tax only in case of war or other extraordinary circumstances.[171] These taxes were then re-used for making war, rather than to provide welfare systems or protect people. Perhaps the only exceptions were poor relief, alms, or specific taxes such as for the repair of the dikes. Many of these special taxes and funds were indirect, however, such as excises on beer, peat, grain, and wine, with fixed sums on consumption. Accordingly, these taxes became a much larger burden on the less fortunate members of society and did not function as a redistribution of wealth from the elites to the poor – actually on the contrary.[172] Similarly, although the 'disaster' tax of the water boards did tax landownership and therefore excluded the poor and landless, it was still felt very unequally. Since

---

[166] Conceded by Scheidel, *The Great Leveler*, 313.
[167] Sellars & Alix-Garcia, 'Labor Scarcity.'
[168] Di Tullio, 'Cooperating in Time of Crisis.' [169] Piketty, *Capital*, 368–375.
[170] Alfani & Di Tullio, *The Lion's Share*, 165–169.
[171] Thoen & Soens, 'The Social and Economic Impact'; Dyer, 'Taxation and Communities.'
[172] Haemers & Ryckbosch, 'A Targeted Public.'

the money had to be paid in a very short time span in cash, it hit the lower-to-middling groups hard, while large landowners could easily oblige.[173]

The same diversity of redistributive outcomes during environmental hazards can be seen during other types of disasters such as famines. On the surface at least, it would appear that famines had the capacity to heighten inequalities more often than to reduce them. Financial buffers ensured that the wealthy could ride out periods when food prices went high, and thus were not forced to sell goods or even land like the poor.[174] However, socio-institutional factors could also conspire to limit this move towards social polarization in times of food crises. Scholars have shown that, rather than complete societal collapse, peasants relied on a combination of reciprocal networks and relationships between individuals and groups at a local level – being offered credit, insurance, charitable sustenance, and capital to ease the burden and avoid having to alienate property as an act of last resort.[175] During some medieval famines, scholars have even suggested a heightened communal sociability – trade switching to restricted local groups based on personal networks of trust and reputation.[176] Sometimes the wealthiest and most powerful segments of pre-modern society had no desire to exploit those further down the social hierarchy because their actual wealth and power was entirely predicated on the maintenance and continued perpetuation of the community as a status quo, thereby satisfying the principles of conservation, defense, and reproduction.[177]

To conclude then, although the views of scholars such as Scheidel and Milanovic about the largely egalitarian effects of terrible shocks such as epidemics appear logical – and are supported in some cases – we still lack a real quantity of empirical examples to fully prove this for the pre-industrial period, or at least to offer this as a definitive universal principle characterizing much of human history. This is not down to a basic lack of interest or material on pre-industrial inequality – a field which has taken off in recent years – but more down to (a) the difficulties of finding source material that can systematically show redistribution outcomes directly connected to a hazard closely before and closely after the shock, (b) the restricted nature of redistributive outcome indicators that are currently used, and (c) the fact that the social groups capable of instrumentalizing the hazard for their own ends were extremely diverse across historical contexts.

[173] Soens, 'Explaining Deficiencies.' [174] Galloway, 'Basic Patterns,' 277.
[175] Vanhaute & Lambrecht, 'Famine.' [176] Slavin, 'Market Failure.'
[177] Di Tullio, *The Wealth of Communities*, 152.

7 Past and Present

Until now, this book has traversed what may be seen as more natural terrain for the historian. To end at this point, however, would be to overlook a crucial component of historical research, especially on a topic like disasters, which is the theoretical and practical contribution to current challenges. This final section therefore poses three questions. Starting from the concept of the Anthropocene, it first of all asks whether the modern period is fundamentally different from the past, and if so, why? Second, it maps the potential of historical research for better understanding vulnerability and resilience to disasters, and equally, the potential of disasters for the study of history. Finally, it outlines some pathways for the future elaboration of disaster history.

7.1 Disaster History and/in the Anthropocene

The profoundly reconditioned interactions between humans and nature in the present age are increasingly perceived as so fundamental as to justify speaking of a new geological epoch – the Anthropocene – in which humankind is fundamentally altering the basic geophysical and biological conditions of life on Earth. The concept of the Anthropocene does away with the modernist distinction between Nature and Society, questions the limits of human agency, and forces us to link the most recent period in the Anthropocene – the postwar period, in which human impact has increased enormously – with the 'deep history' of humankind as predatory species.[1] While there is still discussion on the precise starting point of the Anthropocene on the geological time-scale – the beginnings of agriculture, the Industrial Revolution, the impact of the latter on atmospheric CO_2 via the burning of fossil fuels? – it is clear that the scale of human interactions with the global environment changed dramatically from the 1950s onwards, when almost every indicator of

---

[1] Chakrabarty, 'The Climate of History.'

human enterprise – population, resource use, species extinction, connectivity, etc. – showed an exponential increase.[2]

This Great Acceleration, as it is often called, profoundly altered not only the scale but also the nature and perception of disasters. When Japan was struck on 3 November 2011 by the most powerful earthquake in its recorded history, the country faced a real conundrum of disaster, with 670 kilometers of coast directly affected by a tsunami wave 40 meters high which, apart from destroying or damaging almost one million buildings, also triggered a meltdown of the Fukushima nuclear powerplant.[3] As Sara Pritchard has argued, Fukushima is the ultimate example of the "complex, dynamic, porous and inextricable configuration of nature, technology and politics" in modern disasters.[4] In the "new planetscape of impossibly intertwined entanglings of earthly biorhythms and colossal human engineering projects," John David Ebert argues, the distinction between 'natural' and 'human-made' disasters can no longer be made.[5] Furthermore, a disaster like Fukushima is fundamentally a global one, not only through the evident global media coverage and international solidarity, but also through its impact on nuclear policies, opening a window of opportunity for countries like Germany to shut down their nuclear power plants. The changing attitude towards the potential of technology to prevent disasters is also reflected in the reconstruction policies after the disaster: instead of rebuilding the destroyed houses as soon as possible, Japanese coastal communities hit by the tsunami were resettled on high ground, often invoking a lot of resistance on behalf of villagers that for economic (fishermen) or moral (connection to the ancestral ground) reasons wanted to rebuild their homes in the traditional location near the sea.[6]

There still is debate on whether the Anthropocene itself should be considered as a 'disaster' in its own right – one which probably can only be equaled to the asteroid which about 66 million years ago killed about 70 percent of the species on Earth, including most dinosaurs – or whether there could be such a thing as a 'good Anthropocene,' in which humans and nature will co-evolve into some mutually beneficial 'better' state.[7] More important for our purpose is the question whether disasters in the Anthropocene are indeed profoundly different from any disaster which preceded the Anthropocene, and if so, why exactly? Are the essential changes, if any, situated in the production of disasters, or rather in new types of social vulnerabilities, or shifts in coping

---

[2] Crutzen, 'Geology of Mankind'; Steffen, Crutzen & McNeill, 'The Anthropocene,' 617.
[3] Gill, Steger, & Slater, *Japan Copes with Calamity*.
[4] Pritchard, 'An Envirotechnical Disaster,' 219.   [5] Ebert, 'The Age of Catastrophe,' 4.
[6] Delaney, 'Taking the High Ground,' 63–65.   [7] Ellis, *Anthropocene*, 4.

mechanisms, for instance from risk reduction to resilience? In what follows we consider three features which might set apart the disasters of the Anthropocene: climate change, capitalism, and risk culture.

7.1.1 Climate Change

One debate that has characterized the climate change and disasters literature in recent years is the extent to which climate change – that is, changes in climate resulting from anthropogenic influence – is itself a direct driver of disaster risk. Evidence for the influence of humans on climate has become ever more voluminous. The IPCC currently puts global mean surface temperature in the period 2006–15 at 0.87°C above that of 1850–1900,[8] while model projections suggest that temperatures are locked into a further increase even without continued growth in greenhouse gas emissions. Importantly, this shifting baseline also translates into an increased risk of extreme weather events through changes in their frequency and/or intensity. This can alter the exposure of a society to events such as heat waves, precipitation extremes, and coastal flooding, and consequent overall levels of risk.

Some have argued that climate change has already made an impact on the nature of hazards and disasters. The occurrence of floods and windstorms within the Emergency Events Database (EM-DAT) of international disasters,[9] for example, exhibits a pronounced upward trend over the final decades of the twentieth century, which has led some to argue that we have entered a new age of climate-related disaster. This must be treated with caution, however; data coverage within EM-DAT is poor prior to 1970, and growth since then may be as much a factor of better recording practices as it is a factor of changes in the occurrence or return periods of extreme weather. While this uncertainty is often concluded to be a combination of changes in hazard occurrence, recording, and greater numbers of people and amounts of capital exposed to harm, the relative importance of these factors in producing the apparent increase in disaster occurrence remains unclear.[10]

Similarly, although it has long been recognized that risk and vulnerability to hazards is a "construct of the physical and social worlds,"[11] analyses of the relative roles of the physical and social in producing risk

---

[8] www.ipcc.ch/sr15/chapter/spm/ (last accessed on 26 September 2019).
[9] Hosted by the Centre for Research on the Epidemiology of Disasters at the Université Catholique de Louvain.
[10] Adger & Brooks, 'Does Global Environmental Change Cause Vulnerability?'
[11] Adger & Brooks, 'Does Global Environmental Change Cause Vulnerability?,' 21.

and vulnerability are not consistent across the literature. In mainstream climate change research (i.e. that included within the IPCC Working Group II – Impacts, Adaptation, and Vulnerability), 'risk' – defined as the combination of hazard, exposure, and vulnerability – remains primarily a physical construct, with the degree of environmental exposure and the nature of hazard often overshadowing social vulnerability (Figure 7.1). Furthermore, although the most recent IPCC report notes that vulnerability is multidimensional, it still tends to be conceptualized as a 'second-order' factor; in other words as something that is impacted on by hazards, rather than something that, by way of human agency and deep-rooted social factors, actively shapes the nature of this impact.

These critiques have been most prominent within some quarters of the disaster studies literature and have been brought into focus by the recent media coverage of the contribution of climate change to hurricane

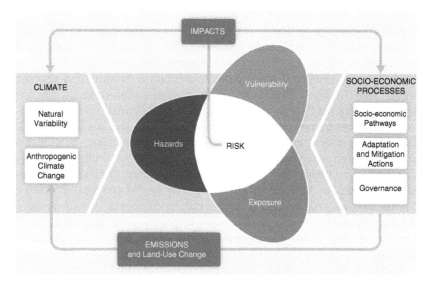

Figure 7.1 Illustration of the core concepts of the WGII AR5. Risk of climate-related impacts results from the interaction of climate-related hazards (including hazardous events and trends) with the vulnerability and exposure of human and natural systems. Changes in both the climate system (left) and socio-economic processes including adaptation and mitigation (right) are drivers of hazards, exposure, and vulnerability. Courtesy of the IPCC. IPCC, 'Summary for Policymakers.'

disasters.[12] Although climate change is thought to influence hurricane intensity through warmer ocean temperatures, Ilan Kelman has argued powerfully that this is largely irrelevant in explaining why the hurricanes of the 2017 season had such significant impact, the issue being rather that a disaster involving a hurricane can happen only if people and infrastructure are vulnerable to it.[13] This vulnerability may arise from a lack of capability or financial capacity to respond effectively to a hurricane, brought about by, for example, lack of stringent building regulations or access to insurance – factors completely independent of climate change. In this view, the debate over whether climate change itself is a driver of disaster therefore represents a return to older arguments concerning the nature of hazard and disaster. More practically, the resultant focus on large-scale efforts to reduce global greenhouse gas emissions may present a 'dangerous distraction' from the more local-level responsibility to implement effective measures to reduce vulnerability.

The longer view nevertheless tells us that both of these perspectives have validity. History is littered with examples of societies in 'marginal' environments where the local-level implications of global temperature change may have pushed the continued viability of human activities in particular environments beyond a certain threshold. It would therefore be unwise to discount the potential for similar changes in the future. On the other hand, what may be perceived as 'unfavorable' longer-term changes in historical climatic conditions did not necessarily lead to economic decline or political instability. On the contrary, historians have argued that Northwest European societies thrived during the cooler climatic conditions of the Little Ice Age, while drier overall Little Ice Age conditions in seventeenth-century Southern Africa appear to have been accompanied by a reduction – rather than an increase – in drought impacts on society.[14] Each of these processes was rooted in levels of social vulnerability rather than in environmental risk.

One can argue, then, that climate change contributes to, and, in many places, increases overall levels of risk by modifying environmental exposure and the nature of hazards. Yet whether this risk translates into disaster is, in the vast majority of cases, determined by society itself. This has two implications for climate change debates. First, reducing underlying vulnerability to present climate variability may not represent a mere 'first

---

[12] This was particularly notable during the 2017 hurricane season, where hurricanes Harvey, Irma, and Maria all had significant impacts on the Caribbean islands and Southern United States.
[13] Kelman *et al.*, 'Learning from the History.'
[14] Degroot, *The Frigid Golden Age*; Hannaford & Nash, 'Climate, History, Society.'

step' towards future climate change adaptation,[15] but may instead hold the key, at least insofar as extreme weather is concerned. In this sense it is unhelpful to cast aside the past as something fundamentally different from the "nonlinear and stepped" changes associated with future climate change,[16] although clearly in areas highly exposed to rises in sea-level, for example, these differences will be felt to a far greater extent. Second, the continued hegemony of research into future physical changes in hazard and exposure at the expense of research into what drives vulnerability may promote a reductionist approach that obscures the critical role of the underlying patterns of vulnerability in producing disasters. Despite many calls for research into such underlying patterns, which may be historically determined, these remain poorly understood in many areas. This represents a major challenge for social scientists and historians to confront.

7.1.2 Capitalism

Underlying the concept of the Anthropocene is the idea that humans or humanity have started to change the physical conditions of the global environment.[17] This might suggest that humanity as a whole can be held responsible for the planetary changes associated with the Anthropocene and the disasters resulting from these changes. But probably, it is more accurate – and fair – to argue that specific humans in specific economic and social configurations were responsible for these changes, while others were forced to cope with the consequences, including increased numbers of extreme meteorological events or technological catastrophes.[18] The best-known spokesman of this position is Jason Moore, who argues that it is not humanity that is responsible for climate change, but capitalism: hence he argues in favor of replacing the concept Anthropocene by *Capitalocene*: the Age of Capital.[19]

The Industrial Revolution is usually portrayed as the main turning point in the relation between humanity and nature. In *The Great Transformation* (1944) Karl Polanyi already meticulously depicted industrial capitalism as a gigantic process of reducing and simplifying land – just like labor – to its mere economic functionality, instead of a vital part of human life, which provided habitation, physical safety, the landscape, and the seasons.[20] Moore, however, retraces the origins of the

---

[15] IPCC, 'Summary for Policymakers.'
[16] As noted by Adger *et al.*, 'Resilience Implications,' 764.
[17] Crutzen, 'Geology of Mankind'; Steffen, Crutzen & McNeill, 'The Anthropocene.'
[18] See for example Ribot, 'Cause and Response.'
[19] Moore, *Anthoprocene or Capitalocene?* [20] Polanyi, *The Great Transformation*, 187.

Capitalocene to the long sixteenth century, when the European-centered modern world economy – as devised by Wallerstein – was taking shape. Around 1450 a turning point was reached, through which humanity's relation with the rest of nature underwent a fundamental change. For Moore, capitalism changed humans' interaction with nature in three ways.[21] First of all, humans and nature were commodified, meaning they could be exchanged and accumulated as labor, food, energy, and raw materials – the 'Four Cheaps' as Moore has labeled them, because of the inherent drive in capitalism of acquiring them as cheaply as possible. By doing so, nature is constantly reworked into a partly human, partly non-human, 'bundle' – a process political ecologists call 'hybridization.'[22] In hybrid form, nature could be mobilized and accumulated. This mobilization over ever longer distances widened the so-called 'Metabolic Rift' between production and consumption – a metaphor of the human body introduced by Karl Marx to analyze the progressive rupture in the nutrient cycle between town and countryside, and later between different parts of the world. Nutrients were extracted in one place, consumed in another, and dumped in a third, hence causing a fundamental socio-environmental disequilibrium and a harbinger of ecological crisis.[23] Abstraction and accumulation are facilitated by a third feature of Capitalism: the 'Cartesian' drive of surveying, identifying, quantifying, classifying, controlling, and sometimes 'protecting' Nature – a logic which, according to Alfred W. Crosby, had already developed into a distinctive feature of European culture and society by the twelfth–thirteenth centuries.[24]

Using a world systemic perspective, Moore argues that the environmental vulnerabilities produced by the Capitalocene are most visible at its margins: the 'frontier' zones of the capitalist system, where cheap resources, labor, energy, and food are found, which can be processed and transferred to the system's core. The 'commodity frontiers' of sugar, cotton, or beef have been mapped as spaces where the new order of the Capitalocene subordinates and in the end erases the old order, but not without exploiting the latter to yield cheap production and unprecedented profits. This is also because the rise of capitalism, in Moore's view, was inextricably linked with colonialism and violent Western expansion, slowly leading to the whole world being incorporated into the capitalist regime.[25]

---

[21] Moore, 'The Capitalocene.' [22] Swyngedouw, 'Circulations and Metabolisms.'
[23] Moore, 'Environmental Crises and the Metabolic Rift.'
[24] Crosby, *The Measure of Reality*.
[25] For a history of cotton from this perspective, see Beckert, *Empire of Cotton*.

In the process, 'frontier' societies often became extremely sensitive to nature-induced as well as technological disasters. The plantation economies that were established in the colonies, based on the Four Cheaps, transformed existing ecosystems and societies, producing not only a different landscape but also new vulnerabilities. In the early-modern Caribbean, for example, the new plantation landscape provided ideal conditions for specific species of mosquitoes, carrying two lethal diseases, yellow fever and malaria, resulting in disease and death among native and slave populations.[26] Something similar happened in the early-modern Southeast of the current-day United States, where the native population was hit not only by continuous slave raids – looking for cheap labor – but also by a transforming disease ecology, as smallpox wreaked havoc among the native population.[27] Colonialism and capitalism not only created new disease ecologies, but according to Davis' well-known work they also turned droughts into famines, causing 'Late Victorian holocausts' from India and China to Brazil. A colonial government unwilling to control the market and focused upon transporting cheap commodities to the homeland did not intervene when harvests failed.[28]

Moore, like Polanyi, in his analysis maintains a rather linear perspective on the development of capitalism – from its medieval localized roots to the world-encompassing system of the present. Scholars like van Bavel, however, recently argued for the existence of capitalist configurations in other contexts as well: for instance in Iraq in the eighth century or China in the Sung period. In each of these contexts, land, labor, and capital became primarily 'processed' and allocated through the market, and in each of these contexts, a dynamic period of economic growth was followed by a period of instability, characterized by rising inequality, collusion between political and economic interests, and mounting environmental problems. From this perspective, capitalism is not the distinctive feature which sets 'modern' history apart from a 'pre-modern past.' Moreover, in each of these contexts, capitalism would itself decline, giving way to a different organization of the economy and the environment, no longer exclusively based upon commodification of production factors.[29] In the future, historical approaches to capitalism combining this idea of cyclicality and a world system analysis might push the analysis one step further.

---

[26] McNeill, *Mosquito Empires*.   [27] Kelton, *Epidemics and Enslavement*.
[28] Davis, *Late Victorian Holocausts*. In a much less controversial way, this of course also links up with the famous work of Amartya Sen, on the Bengal Famine in 1943. Sen, however, sees 'democracy' as the main solution for defying hunger and does not question the foundations of global capitalism.
[29] Van Bavel, *The Invisible Hand?*

7.1.3 The Risk Society

Writing in the 1980s, Ulrich Beck formulated the idea of the present age as a 'Risk Society,' a society in which disasters stopped being exceptions, a break from daily routines, and had become part of everyday life. For individuals, for communities, and for states, dealing with the risk of hazards and disasters became a central preoccupation. In other words, people started to live in constant fear.[30] In the wake of Chernobyl and Bhopal, the disasters Beck had in mind were predominantly technological or, more precisely, they were hybrid configurations of nature and technology. As hazards and disasters could no longer be avoided, resilience – bouncing back and adapting – gradually replaced vulnerability as the dominant framework in disaster analysis and policy.[31]

In the Risk Society, natural disaster is increasingly framed as inevitable. Quite paradoxically, however, much of the present-day vulnerability to natural disaster resulted from the ambition to control nature, using technology, creativity, and capital. In this respect the antecedents of the modern Age of Risk can be situated much earlier. In the eighteenth and nineteenth centuries, the Enlightened naturalization of the world definitely set society and nature at different poles: nature became something 'out there,' waiting to be understood, controlled, and conquered. Meanwhile, the Industrial Revolution greatly expanded the technological possibilities allowing one to succeed in this conquest of nature. As a result, nature-induced disasters were increasingly presented as failures of control, calling for greater human endeavors to avoid their repetition. In this respect, the Lisbon Earthquake of 1755 is often considered a turning point in (Western) dealings with risk, for a number of reasons. Among the reasons are the degree of central coordination by the state shown in the recovery from the earthquake, the efforts of this state to rationalize, measure, calculate, and undo its impact, and the wide range of technological improvements deployed to make the built environment more resilient to future earthquakes. Lisbon, however, was also one of the first disasters which was widely discussed in an emerging public sphere of newspaper-readers and intellectuals, all over Europe and the colonial world.[32] The earthquake made a deep impression on the intellectual world of the Enlightenment, with its adherents such as Voltaire and Kant publishing extensively on the subject and each in their way contributing to both the scientific study of disaster and its naturalization, with Voltaire in both *Candide* and the *Poème sur le désastre de Lisbonne*

---

[30] See Section 2.3.6.   [31] See Section 7.2.3.
[32] Koopmans, 'The 1755 Lisbon Earthquake,' 26–29.

vehemently attacking those who still believed that such disasters were some form of divine punishment.[33]

If disasters were natural, blind, and evil, human industry should and could be directed at preventing their occurrence and controlling their impact. And in the two centuries following the Lisbon Earthquake, the technological possibilities to do so expanded greatly, and huge numbers of people started to settle in flood-prone deltas or practice irrigation-farming in water-poor regions, as if floods and droughts did not exist. The coming of age of the Anthropocene replaced the ideal of absolute safety by the ideal of acceptable risk – a relative degree of safety based on accurate calculation and assessment, permanent alert, smart use of technology, and maximal accommodation of hazards.[34] The idea of 'acceptable risk' is very prominent, for instance in modern coastal engineering. On the basis of projections of the frequency and intensity of extreme sea levels in the future, as well as calculations of relevant uncertainties, flood protection is continuously being adapted in order to withstand 'once-in-a-thousand-years' or even 'once-in-ten-thousand-years' storms.[35] However, if 'acceptable risk' is one side of the coin, the fundamental unpredictability of modern disasters is the other. Disasters like 9/11, Chernobyl, or the 2004 Indian Ocean Tsunami by far exceeded the margins of probability of commonly available risk assessments, leading Joachim Radkau to the suggestion that only science fiction and horror stories could provide realistic scenarios for some of the disasters unfolding in the modern age of risk.[36]

In sum, it is clear that anthropogenic climate change as well as the rise of the capitalist world-system, or the risk society, drastically altered the production, the impact, and the handling of disasters. At the same time, however, the gap between Anthropocene disasters and disasters in the more distant past is often surprisingly small, especially when discussing vulnerability and resilience. Especially the roles of different institutional formations and coordination systems (state, market, family) in relation to hazards, and those of social actors and their sometimes differing interests, can be instructive with respect to present-day situations. The many examples and case studies discussed throughout this volume make clear that the study of past disasters, even those which occurred in the distant past, can offer a substantial contribution to a better understanding of disasters today.

---

[33] Hamblyn, 'Notes from the Underground.'
[34] Knowles, 'Learning from Disaster?,' 778.
[35] Wahl et al., 'Understanding Extreme Sea Levels.'
[36] Radkau, *Nature and Power*, 265–271.

7.2 The Potential of History for Better Understanding Disasters

Of the many fields that contribute to mainstream disasters discourse, history is often found towards the bottom of the list.[37] In much of this literature we might find a brief preamble on the history of a particular disaster or policy over some decades – the so-called 'long term' – while the present tends to remain rigidly detached from the past, creating an artificial divide between knowledge perceived as relevant and that seen as irrelevant for disaster risk reduction. Equally, we might argue that much scholarship focusing on historical disasters remains detached from the present. Many arguments for historically informed disaster research have come not from historians, but from geographers, sociologists, ecologists, and paleoclimatologists – a circumstance that has shaped the ways in which scholars have attempted to draw 'lessons' from the past.

One of these is known as 'forecasting by analogy' – an approach pioneered by social scientist Michael H. Glantz in the late 1980s and 1990s. This approach views past experiences of responses to hazards as analogous to future challenges, arguing that, if we can identify how past societies successfully or unsuccessfully managed risk, then we can use this knowledge to forecast the likely impacts of future threats such as climate change.[38] Past disasters are turned into 'completed natural experiments of history', which can be mobilized to forecast the range of potential outcomes of future disasters (Figure 7.2).

In the field of disasters, analogy-based methodologies grew in popularity during the 2000s as the rapid growth in paleoclimate proxy data began to shed greater light on past climatic change, which in turn facilitated an increasing number of studies that zoomed in on episodes of societal 'collapse' in past millennia that coincided with episodes of significant climatic change.[39] Yet it was largely because of an explicit focus on discrete periods of abrupt environmental change and collapse that these analyses have been left open to criticisms of determinism and oversimplification. In particular, analogies have been criticized for reducing the

---

[37] We can broadly define 'mainstream' as that research discussed by the United Nations International Strategy for Disaster Reduction (UNISDR), or in journals such as *Disasters*, the *International Journal of Mass Emergencies and Disasters*, or the *International Journal of Risk Reduction*. A historical perspective also remained absent from influential handbooks of disaster research, such as Rodríguez, Quarantelli & Dynes, *Handbook of Disaster Research*.

[38] See Glantz, *Societal Responses*; Glantz, 'Does History Have a Future?'; Glantz, 'The Use of Analogies.'

[39] For examples, see Hodell, Curtis & Brenner, 'Possible Role of Climate'; Dugmore *et al.*, 'Cultural Adaptation'; Holmgren & Öberg, 'Climate Change'; Riede, 'Towards a Science of Past Disasters.'

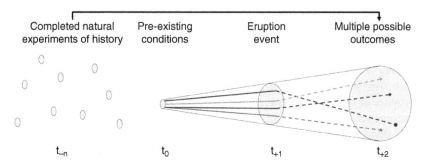

Figure 7.2 So-called 'scenario trumpet' projecting possible disaster scenarios, building on experiences of the past. Riede, 'Past-Forwarding Ancient Calamities.'

societies in question to respondents to a series of exogenous threats, for uncritically drawing lessons from societies that are markedly different from those of today, and for the lack of a suitable analogy for contemporary climate change.[40] Despite such critiques, analogies retain considerable currency, particularly as communication tools.

It was off the back of these criticisms that the Integrated History and Future of People on Earth (IHOPE) network emerged.[41] This approach derives from social-ecological systems analysis, which views humans and the environment as one holistic system that is defined by its level of resilience to disturbance.[42] This school of thought rejects the notion of past, present, and future as separate entities and instead conceptualizes temporality as the 'long now.'[43] In doing so, it integrates historical (largely archaeological) data into systems models to identify 'safe and just' spaces and boundaries for humanity to operate within – its ultimate aim being to provide recommendations to build sustainability.[44] A focus on systems rather than people and a reliance on archaeological rather than historical evidence has nevertheless left IHOPE exposed to some of the same criticisms directed at analogies, in that they tend to present historical trajectories without recourse to the human agency and uneven

---

[40] Adamson, Rohland & Hannaford, 'Re-thinking the Present.'
[41] IHOPE was founded in 2003 by the ecological economist Robert Costanza, see Costanza et al., 'Sustainability or Collapse'; Costanza, Graumlich & Steffen (eds.), *Sustainability or Collapse?*
[42] Berkes & Folke (eds.), *Linking Social and Ecological Systems*; Holling, 'Understanding the Complexity'; Folke, 'Resilience.'
[43] Dearing et al., 'Safe and Just Operating Spaces'; Redman & Kinzig, 'Resilience of Past Landscapes.'
[44] Dearing et al., 'Safe and Just Operating Spaces'; Rockström et al., 'Planetary Boundaries.'

7.2 The Potential of History

distributions of power embedded within decision-making, and are therefore ultimately reductionist.[45]

It is all very well for historians to label these approaches as 'deterministic' or 'reductionist,' but are there historical approaches that can counteract these criticisms? Indeed, historians have been relatively slow to make the case for historical disaster research that moves beyond the view of the past as an end in itself, and instead to see history as a vehicle to understand how societies engender, respond to, and recover from disasters.[46] This is in part due to a fear of appearing deterministic. Even historical climatologists have largely avoided connecting their research to the present until relatively recently, despite making use of frameworks derived from the contemporary global climate change literature such as vulnerability and resilience. This is unfortunate given that the richness of the historical record allows us to reconstruct the social, economic, and cultural impact of hazards and shocks over time periods simply not possible in contemporary disaster studies material – allowing us to better observe structural societal changes rather than short-term immediate disruption that may be rectified within a matter of years.[47] Only recently have calls for historically informed disasters research begun to emerge from historians themselves. We now review some of the arguments which these calls have made.

7.2.1   The Historical Roots of Present-Day Disasters

Throughout this volume, it has been shown that contemporary disasters sometimes had deep roots in historical processes: the impact of the terrible earthquake in Peru in 1970, for example, has been linked to Spanish colonial practices over 500 years ago.[48] Extreme destruction and suffering turned out to be as much products of Peru's long history of underdevelopment as they were of the earthquake itself. Urbanization patterns – with their dense concentrations of humans, buildings, and capital – were the long-term remnants of Spanish subjugation of the indigenous population in the sixteenth century. Spanish colonizers forcibly moved rural peasants out of their traditional dispersed habitation patterns, where they had found a complex socio-economic balance in this difficult environment and were able to spread and buffer risks, into these central agglomerations. The collective institutions (formal or informal), redistributive systems, and personal reciprocity required to reduce the

---

[45] Adamson, Rohland & Hannaford, 'Re-thinking the Present.'
[46] Curtis, van Bavel & Soens, 'History and the Social Sciences.'
[47] Van Bavel & Curtis, 'Better Understanding Disasters.'
[48] Oliver-Smith, 'Peru's Five-Hundred-Year Earthquake.'

impact of severe shocks had long been eroded by the installation of *hacienda*-based modes of exploitation worked by mass ranks of marginalized Indian serfs. Similarly, the devastating earthquake that struck Haiti in 2010, killing approximately 250,000 people, has been linked to a preexisting state of societal vulnerability inherited from very particular paths taken far back at the beginning of the nineteenth century.[49] A skewed relationship between state and society, epitomized by frequent patterns of violence and peasant resistance and revolt, led to a vicious circle of reduced rent extraction opportunities, a weak or 'failed' state, and ultimately a dearth of growth-enhancing and protective public and private institutions – the absence of which was keenly felt when the earthquake occurred in 2010.

Some would conclude that such path dependencies can render us 'prisoners of history,' as institutional arrangements shaping human interaction with hazards tend to embody past understandings and imperatives rather than those attuned to the present.[50] Institutions are deeply embedded in the societal context within which they were formed, and are shaped according to power dynamics and the memory of issues they have addressed over the course of their existence. This historical context is frequently hidden from view in the disasters literature, and as argued there is an urgent need for new lines of research that trace how institutions have functioned historically in relation to hazards, and that map out where and how path dependencies become active over time.[51] Similar calls have been made in the climate change adaptation literature, although these have been criticized for the relatively shallow time depth that is often employed.[52] Indeed, in path-dependent patterns, formative moments or critical junctures drive or reinforce divergent paths. Having crystallized into institutions – including cultural values – these paths become difficult to shift. Increasing costs develop over time when switching from one policy alternative to another, making future change and adaptation more difficult.[53] These deep causes become especially highlighted when faced with new circumstances brought on by exogenous shocks.[54] In order to uncover path-dependent processes, then, one must begin from a critical juncture that underlies subsequent events, which necessitates historical research. The identification of path dependency can therefore allow one to acquire better understanding of the long-term drivers of vulnerability, and ultimately lead to better-targeted

---

[49] Frankema & Masé, 'An Island Drifting Apart.'
[50] Dovers & Hezri, 'Institutions and Policy Processes.'
[51] For this argument, see Section 5.3.2.   [52] Wise *et al.*, 'Reconceptualising Adaptation.'
[53] Pierson, 'Increasing Returns'; Elvin, 'Three Thousand Years of Unsustainable Growth.'
[54] Dietz, Stern & Rycroft, 'Definitions of Conflict'; 't Hart, 'Symbols.'

interventions that avoid potential unintended consequences that can arise in the absence of sensitivity to historical and social contexts.

7.2.2 The Past as an Empirical Laboratory: Institutions and Social Context

One key argument underpinning this monograph is that 'the past' can be used as a 'laboratory' to empirically test hypotheses of relevance to the present, by spatially and chronologically comparing the drivers of and constraints on societal responses to shocks – in turn enriching our understanding of responses to stress today. This approach sees history as a science: it moves beyond 'the narrative' and 'the particular,' and seeks to identify distinct or divergent patterns, constellations, and trajectories. This can help guard against teleological forms of explanation, or explanations following uni-linear forms of progression to an inevitable outcome.

Despite arguing for a focus on patterns and trajectories, we also advocate for a focus on social context and its role in shaping responses to hazards in particular regions and localities. First, hazards and human responses to them have regionally and locally specific characteristics: even 'global' phenomena such as climate change are experienced in the form of local processes such as coastal erosion or various forms of extreme weather, with sometimes very differing social consequences, and therefore no singular approach to reducing risk and vulnerability fits neatly across all contexts. As a result, it was often the 'export' of institutional and technological 'solutions' to hazards and disasters from one region to another which proved particularly problematic, as such solutions not only failed to do what they were meant to do, but also created new types of problems – sometimes directly paving the road to future disaster themselves.[55] Similarly, the global climate environment may 'drive' epidemic disease outbreaks, but they represent only the larger framework within which local contextual environmental and societal conditions dictate actual epidemiologic outcomes – pathogens move to human hosts under contextually specific conditions.[56] This contextualized view is nevertheless sometimes lacking in disasters discourse. Equally, the hegemony of model-based approaches in the IHOPE school, for example, can overlook the social and cultural attributes of a region. The kind of 'within region' systematic comparative approaches that we have made the case for can therefore be accompanied by detailed, long-term analyses of human interaction with the 'inbuilt' hazards of a particular place, which can ensure that responses are grounded within their place-specific context.[57] This can also

---

[55] See Sections 4.2.1 and 5.2.2 for examples. [56] Brook, 'Differential Effects.'
[57] See Section 3.2.3.

extend to an investigation of how past hazards and disasters become inscribed in the cultural memory of a region, which has been shown in a number of contexts to inform responses to hazards today.[58]

Building on this appreciation of social context, a second element calls for a more contextualized approach to historical disaster research that focuses on the evolution and functioning of institutions. Responses to hazards and disasters always take place within the broader context of institutions – be they formal or informal – which play a crucial role in driving or constraining vulnerability. By this, we do not mean only those institutions that are specifically set up to deal with hazards, which much of the disasters literature focuses on exclusively, but also those 'indirect' institutions that exist regardless of the presence of hazards.[59] While it is common in the disasters literature to compare various institutional responses against disaster outcomes across a diverse range of contexts, whether that be at the market, state, communal, or household level, recent work has questioned the validity of this exercise, instead noting that types of institutional arrangements do not *per se* have an intrinsic value in reducing risk, but only 'make sense' by being embedded within their social settings.[60] This is significant, given that a varied array of research in the disciplines of history, sociology, and political economics tells us that institutions do not always evolve towards a state that promotes an optimum level of societal resilience, but rather tend to drift towards the needs of restricted interest groups, especially those with the most bargaining power and access to resources.[61] Institutions necessary for welfare and protection may also have their performance and functions inadvertently affected by interaction with other institutions.[62] One further process that historical research can illuminate, then, is how the deep-rooted interests and preferences of certain social groups who control the functioning of institutions may dictate their effectiveness in dealing with hazards in different contexts. This can, in turn, help ensure that responses to hazards do not simply reproduce inequitable power structures and create self-reinforcing cycles of vulnerability. In this book, we have elucidated ways we can approach these lines of investigation most fruitfully through systematic comparative analysis.[63]

---

[58] Endfield, 'Exploring Particularity'; Endfield & Veale (eds.), *Cultural Histories*. See also Sections 4.5 and 5.2.1.
[59] Van Bavel & Curtis, 'Better Understanding Disasters.'
[60] Van Bavel, Curtis & Soens, 'Economic Inequality.'
[61] Ogilvie, '"Whatever Is, Is Right"?'
[62] Ogilvie & Carus, 'Institutions and Economic Growth.'   [63] Especially in Section 2.3.3.

7.2.3 The Great Escape: Can History Teach Us How to Escape from Disaster?

Disaster history is generally considered a dark and gloomy field of history, telling stories of suffering and despair, of vulnerable people whose livelihoods were suddenly and brutally disrupted. However, history also hides many examples of regions and societies which once have been vulnerable to a particular threat, but where this threat has eventually been overcome, or at least strongly contained. Careful historical analysis might reveal the conditions and drivers that produced such an 'escape from disaster.' A classic, though highly disputed, example is to be found in the history of famine. After the 1845–47 potato famine, harvest failures in Europe no longer induced massive starvation (which became confined to contexts of warfare and to people experiencing 'marginal' living conditions).[64] Moreover, in some regions such as Holland, this 'escape from famine' may have been achieved already at a much earlier stage, in the course of the sixteenth or seventeenth century.[65] Hence it becomes tempting to frame such a retreat of vulnerability as a by-product of 'modern' economic growth, producing the technological advances necessary to remove the Malthusian limits on food production, the economies of scale and levels of market integration needed to overcome regional food shortages through trade, and the gains in productivity that made it possible to reduce the importance of food in the expenditure of the household.

If poverty was the ultimate cause of disaster vulnerability, then growth in welfare provision might be the solution. More generally, economic growth may be seen as the crucial factor in reducing vulnerabilities, including those related to natural hazards. But can we indeed observe such positive interaction between economic growth and reduced vulnerability throughout history, and if so, what were the underlying mechanisms explaining the positive impact of economic growth on disaster vulnerability? A way of approaching this question, focusing on the relation between economic growth and environmental problems in general, is offered by the so-called Environmental Kuznets Curve (EKC).[66] In parallel to the relationship between economic growth and social inequality put forward by Simon Kuznets, the EKC or 'inverted U-curve' predicts that economic growth will initially produce increasing environmental problems. When the growth becomes sustainable, however, the environmental impact will stabilize and perhaps even decrease again. Because pollution can be measured in a relatively uniform way, the

---

[64] Ó Gráda, *Famine*, 36; Fogel, *The Escape from Hunger*.
[65] Curtis & Dijkman, 'The Escape from Famine.'
[66] Grossman & Krueger, 'Economic Growth'; Klein Goldewijk, 'Environmental Quality.'

model is mostly used in pollution studies, although it has been applied to disaster impact as well.[67] The basic causal mechanism behind the stabilization of environmental problems where prolonged growth exists is the demand for a 'clean' environment (or better protection against disasters), which is believed to increase in parallel to income and standards of living (and which becomes an issue once a significant proportion of the population is no longer predominantly occupied with fulfilling needs of subsistence). Meeting the requirements of a safer and cleaner environment, while maintaining higher levels of income per capita, becomes possible as efficiency gains make it possible to release capital which can be invested in environmental protection. In the sphere of human health, the obvious gains in life expectancy and diminished exposure to epidemic diseases and mass mortality in developed countries might be linked to improve standards of hygienic, better nutrition, and better health care. All of this could be afforded by higher income. And indeed, in the twentieth century, there was a fairly stable positive correlation between life expectancy at birth and GDP per capita, with richer countries indeed witnessing higher life expectancies. Interestingly, this was not yet the case in the nineteenth century, when countries with a higher per capita income might even have seen lower life expectancies, compared with 'poorer' countries.[68] While this suggests the existence of an environmental Kuznets curve, of course, aggregate data on life expectancy and exposure to disease provide no information on the important social differentiations in health care which continue to exist in developed economies, and might even become stronger with the increase in inequality, and/or reductions in public expenditure on health care.[69]

In the past, sustained economic growth provided the funding to carry out big engineering projects, like the major improvements in flood protection in the North Sea area after the 1953 floods. These improvements coincided with the postwar economic boom, and helped to reduce the theoretical likelihood of major dike breaches from once in 100 years to once in 10,000 years. We have also seen, however, that technological solutions can create either a false sense of security or dangerous side-effects and thus increase rather than decrease vulnerability.[70] In the history of agriculture, economic growth might enable higher capital inputs (mechanization, fertilization) as well as investments in research and development, all working to reduce the potential of adverse weather to seriously disrupt the harvests. On the other hand, new 'high-yielding'

---

[67] Cavallo *et al.*, 'Catastrophic Natural Disasters'; Kellenberg & Mobarak, 'Does Rising Income.'
[68] Zijdeman & Ribeiro da Silva, 'Life Expectancy,' 112.   [69] See Section 4.4.
[70] See Section 5.2.2.

crop varieties are often highly demanding in technical terms, as well as being vulnerable to distortions in both inputs and weather. In the end, advanced agriculture remains as dependent on weather and diseases as its 'traditional' predecessors; only the effects of a harvest failure might be different.[71] Or, as William H. McNeill observed in 1992 in what he termed the 'conservation of catastrophe': "It certainly seems as though every gain in precision in the coordination of human activity and every heightening of efficiency in production were matched by a new vulnerability to breakdown."[72] Hence, the existence of an inverted U-shape in the relation between economic growth and the occurrence or impact of disasters remains to be tested.

7.3 The Potential of Disasters for Historical Research

While historical research can contribute to understanding disasters, the reverse is also true: studying past disasters can enrich historical research. Historians are by no means unfamiliar with disasters; they have been writing about catastrophes for a long time. In fact, few historians will object to the observation that disasters have played a central role in many cultures. Fear of hunger and famine often had a pervasive impact on the organization of food production in rural communities, explaining the so-called 'prudence of the peasants.'[73] The successful management of natural hazards often became a cornerstone of political power: the Chinese imperial constitution, for instance, turned disasters into serious challenges for the emperors: the Mandate of Heaven saw the emperors as the ultimate connection between Heaven and Earth, and uncontrolled natural disasters might indicate that the emperor had forsaken this mandate.[74] Origin myths also often start from mega-disasters. The enduring importance of Noah's Flood for Jewish, Christian, and Islamic cultures is probably the best-known example – it even proved very inspiring for the development of geology, as the biblical catastrophe could apparently explain the occurrence of marine sediments and fossils of sea animals high above sea level, an interpretation known as 'Neptunism.'[75] And after all, with the idea of the Big Bang, modern science is still arguing for a disaster as starting point – and potential end – of our universe: earth's origins remain rooted in catastrophe. In national histories as well,

---

[71] Federico, *Feeding the World*, 12. In the United States the average variability of wheat yields was higher in the period 1960–2000 than in the period 1860–1910.
[72] McNeill, *The Global Condition*, 148, cited by Mauelshagen, 'Defining Catastrophes,' 183.
[73] McCloskey, 'The Prudent Peasant,' see above Section 5.2.2.
[74] Brook, *The Troubled Empire*.
[75] Bowler & Rhys Morus, *Making Modern Science*, 111–118.

disasters frequently occupy a prominent place. Societies often developed an obsession for particular types of disaster, and in some cases these disasters turned into a cornerstone of national identity: as we have seen above, this was the case for famine in Ireland, for floods in the Netherlands, and for disasters induced by colonialism in many parts of the Global South.[76]

Throughout this book we have highlighted two ways in which more attention to disasters might benefit our understanding of history. On the one hand, disasters sometimes turned out to be 'historical protagonists,'[77] forcing, accelerating or facilitating changes in the economic, cultural, social, or political organization of society. On the other hand, disasters might also reveal features of societies which remain hidden in 'normal' situations, but become exposed in times of crisis. These themes are now consolidated, before exploring future pathways for disaster history.

7.3.1 Disasters as Historical Protagonists

As we have seen, historians have long been reluctant to ascribe too much causality to disasters, in particular to nature-related disasters. Throughout much of the twentieth century, environmental determinism was considered outdated, a relic of the past. Because of their apparently random and insignificant nature, nature-induced disasters were considered unlikely to cause anything like a long-term structural impact. Back in the 1960s, even Emmanuel Le Roy Ladurie, the pioneer of Climate History, considered the 'human' effects of climate extremes almost irrelevant.[78] As late as 1980, Jan de Vries famously argued that "short-term climatic crises stand in relation to economic history as bank robberies to the history of banking."[79] Opinions were about to change, however. As early as 1989, Mark Overton challenged the claim by de Vries, saying that bank robberies can still be proximate causes of changes in the banking system; structural responses could outstrip the economic effect of the individual bank robbery in question.[80] But, especially since 2000, a rapidly expanding body of historical literature has argued for a more active role for disasters, not only in the short term, but also in the long term, as vectors of lasting, structural changes.

As we have seen in Section 6.3, shocks and disasters are increasingly being rediscovered as the 'missing link' in the explanation of major dynastic, demographic, or economic crises, or the collapse of entire

---

[76] See above, Section 1.2. [77] Campbell, 'Nature as Historical Protagonist.'
[78] See above, Section 1.2. [79] De Vries, 'Measuring the Impact,' 603.
[80] Overton, 'Weather and Agricultural Change,' 77.

civilizations. Well-known disasters like the Lisbon Earthquake or the fourteenth-century Black Death, but also hitherto unknown or poorly studied 'mega-droughts' and pandemics, are being framed as important drivers of change. Sometimes the disaster or shock is identified as 'the prime mover' of an important transition, while on other occasions it is seen mostly as an accelerator or catalyst for processes that were already unfolding. In the latter case the disaster might have opened a 'window of opportunity' allowing specific actors to finally impose a program of change they already had in mind. As we have argued, history can still gain a lot by paying more attention to the precise nature of the interactions between 'exogenous' shocks and 'endogenous' features of the societies affected by the disaster, notably by developing more systematic spatial and chronological comparisons, making it possible to disentangle the disaster from its context as much as possible.[81]

7.3.2 Disasters as Tests at the Extreme Margin

There is yet another way in which disaster history can directly inspire other fields of history. As we have seen, the significance of disasters did not remain constant, but was profoundly different from period to period, and from region to region. One can know a society through its disasters – so to speak. Disasters put societies under pressure: they are tests at the extreme margin. As such, they bring to light latent qualities and characteristics of societies, features that under normal conditions do not stand out.

Among those features, inequality is an important element. As demonstrated in Chapters 5 and 6, the vulnerability of groups and individuals is often closely linked to their social, economic, and political position in society. People at the margins of society – lacking capital, political influence, or cultural status – tend to be the most exposed to the negative impact of disasters. In turn, disasters may further erode the resources of victims, thus reinforcing and exacerbating existing inequalities. Studying past disasters can reveal aspects of inequality that otherwise would have remained hidden from view. The differential impact of shocks, in particular, may show which groups lived closest to the edge, how precarious their position was, and frequently also which underlying mechanisms explained their vulnerability. That New Orleans was divided along racial and socio-economic lines was well known even before Hurricane Katrina, but exactly how deep the divisions were became crystal clear during and after the storm. Low-income African-Americans in particular were hit

---

[81] Curtis, van Bavel & Soens, 'History and the Social Sciences,' 761–765.

harder than other groups: they were over-represented among the group that did not manage to leave the city before the storm struck and they were far more likely to lose their job afterwards. African-Americans in general, poor or not, reported higher levels of stress in the aftermath of Katrina than whites.[82]

In the history of colonialism and imperialism as well, disasters have often been scrutinized to reveal the underlying systems of economic exploitation and political dominance. For Mike Davis, only the massive starvation following *El Niño*-related droughts on the Indian subcontinent from the 1870s onwards can accurately capture the economic logic of the imperial world-system, with its forced concentration on massive exports of grain and cotton, and destruction of the traditional resources of resilience.[83] Floods have also been studied to uncover the nature of colonial rule. In the period 1880–1950 the British colonial government made important efforts to regulate the rivers of Northern India, for purposes of flood safety, transportation, and irrigation.[84] The construction of large-scale embankments prevented alluvial deposit formation, impeded drainage, fostered the spread of malaria, and in the end increased rather than decreased the flood risk. In the analysis of Rohan D'Souza, dam building became the ultimate instrument of colonial capitalism, ending centuries-old flood-dependent agrarian regimes and installing an export-oriented agriculture based on a private property regime.[85]

Social and political inequalities are not the only features of societies exposed by disasters; the institutional framework is another. Here, perhaps, the role of disasters as tests at the extreme margin becomes clearest. The capacity of societies, groups, and individuals to cope with shocks depends partly on institutions: the rules, customs, practices, and organizational forms that shape the response to shocks. Studying the response of historical societies to shocks can, first of all, reveal the robustness of institutions under adverse circumstances. Research has, for instance, shown that in late-seventeenth- and early-eighteenth-century France grain markets did not, as was previously thought, 'balkanize' during famines through local imposition of restrictions on food shipments. Regional and interregional grain trade, with its emphasis on the provisioning of Paris, continued to operate largely as it did in normal years. Thus the organization of the grain market proved highly 'robust,'

[82] Elliott & Pais, 'Race, Class, and Hurricane Katrina'; Hartman & Squires (eds.), *There Is No Such Thing*.
[83] Davis, *Late Victorian Holocausts*.
[84] Hill, *River of Sorrow*; Singh, 'The Colonial State.'
[85] D'Souza, *Drowned and Dammed*. See also Section 7.1.2.

although this also implied prioritizing the capital over provincial cities and the countryside.[86] This observation leads us to a second question: did institutions operate in ways that allowed people to cope with shocks? Did they, in other words, reinforce the resilience of society as a whole, and of vulnerable groups in particular? Whether grain markets in early-modern France had this effect is doubtful. Not only was mortality very high – the famines of 1693–94 and 1709–10 together killed more than two million people – but also urban demand, backed by superior urban purchasing power, inflated prices in the production areas disproportionately: in the Paris basin the volatility of wheat prices was higher than in the city itself.[87] Through the history of these famines, a better understanding of both the French grain market and the degree of political centralization can be achieved.

Finally, disasters test the capacity of societies to learn and adapt in order to prevent recurrences, or at the very least to mitigate the impact of subsequent shocks. Returning to Hurricane Katrina, responses to the storm were thoroughly examined afterwards in a series of official reports. However, these reports focused largely on the actions – or absence of actions – by various governmental agencies at the federal, state, and local level and on the need for adjustments of communication systems and technological infrastructure. The reports paid virtually no attention to issues such as poverty and race, which had been identified as root causes of vulnerability in academic research.[88] The case of Katrina thus demonstrates how the capacity of societies to learn from disaster can be restricted by political and social biases. Disasters, in short, act as magnifying lenses which expose aspects of past societies that might otherwise have escaped the eye.

7.4   Future Pathways

It is difficult to predict what the future will hold for the field of disaster history, but certain trends will most probably unfold in the coming years. One of these is interdisciplinary research. The field of disaster history is by definition crowded with natural, social, and human scientists; nevertheless, true interdisciplinary research is still quite rare, and the chasm between these fields is seldom bridged. The niche of climate history may be the exception to the rule. Here, both climatologists and historians have created long-term climate reconstructions based on a wide range of

---

[86] Ó Gráda & Chevet, 'Famine and Market.'   [87] Meuvret, 'Les oscillations des prix.'
[88] A Failure of Initiative; Lessons Learned; A Nation Still Unprepared (all available from www.disastersrus.org/katrina).

data, archival sources, and methodologies. Most impressive, however, is the tendency of the social, human, and exact sciences to approach these interdisciplinary sources and datasets not from the isolated perspective of their individual discipline, but rather bringing an open mind to tackle questions and to perform analyses from different disciplines. Historians have stepped outside of the realm of the human sphere and looked at ecosystems, weather patterns, and climatic shifts in their own right, while ecologists and climatologists have been interested in the societal impact of climatic influences. In this way, both sides of the scientific world have cross-fertilized one another.[89]

This interdisciplinary success should have a role as an exemplar for the other sub-fields of disaster studies. Historians should embrace the methodologies, data, and findings of the other social sciences as well as those of the exact sciences to get a better understanding of disasters in the past and present. This should not have to lead to an adaptation to or appropriation of the perspectives and methods of these different disciplines, however. Interdisciplinary research should be a two-way interaction whereby the disciplines involved could work on a common set of concepts, topics, and questions that can be approached in a truly interdisciplinary way, rather than merely correlating or combining multiple independent datasets and methodologies.

Importantly, and perhaps obviously, historians can convey the importance of time and chronology in studying disasters. As this book has shown, general patterns of behavior do occur in different places and in different time frames. Some patterns have a linear temporal distribution, while others have a cyclical or sporadic recurrence. For some types of disasters, there are clear precedents in recent history, for others only a perspective spanning several centuries can inform us on causes, vulnerabilities, and impact, and yet other categories of disaster – think of volcanic eruptions in Western Europe – require an archaeological or geological time-scale, covering at least several millennia.[90] In the analysis of disaster responses, effects, and preconditions as well, a historical approach problematizes a purely processual interpretation of disasters, vulnerability, and adaptability. Context always mattered, and responses which were successful in one context utterly failed in others. In unraveling the intimate relationship between the disaster and its particular social, temporal, and geographical context, historians can offer a major contribution to the study of disasters.

---

[89] Allan *et al.*, 'Toward Integrated Historical Climate Research.'
[90] Riede, 'Past-Forwarding Ancient Calamities' on the Laacher See volcanic eruption in Western Germany, about 13,000 years ago.

7.4 Future Pathways

This brings us to a second pathway: history as a laboratory. The past provides us with a wide range of disasters set in different time frames, and more especially in different types of societies, with very specific socio-economic, political, and cultural structures. As a result, historians can test how different social, economic, cultural, and political constellations affect the adaptability and vulnerability of societies or social groups. By comparing and analyzing these differences and similarities, historians can distinguish general patterns and trajectories that occur in distinct societal circumstances. This book has attempted to provide such an approach. It is, however, important to apply this approach beyond the examples that we have considered here and increase the amount of comparative research to broaden this perspective. A future goal should be to look for the most basic causes of social vulnerability and adaptability. What causes societies, and especially social groups, to become vulnerable to natural hazards? Which institutions, social structures, political actors, and cultural settings provide the best options – depending on the specific context – to respond efficiently to crisis? Only when certain patterns and trajectories become clearer can historians move beyond the narrative and the particular and provide assessments that may even inform policy.

A third pathway for the future is to challenge the Eurocentric approach to disaster history that has predominated, at least until relatively recently. A first challenge is to expand the amount of research concerning non-Western societies and their mechanisms for coping with natural hazards and shocks in the past and present. Unfortunately, most of our current overview is highly focused on European or Western contexts, because these regions and societies are highly over-represented in the available literature. As Greg Bankoff has pointed out convincingly, Western perspectives on risk, mitigation, danger, and vulnerability still dominate how we look at regions and disasters.[91] As a result, the Global South is often portrayed as unsafe, highly vulnerable, and less resilient. In addition, researchers often look for Western institutions, responses, and mitigation measures across the globe, instead of analyzing the alternative structures and strategies in their own right. The centralized and highly technological prevention, response, and mitigation measures that are common in Western countries have been set as a standard against which all other disaster prevention and mitigation measures are tested. As our analysis has shown, however, even within Europe a plethora of responses existed. There never was a single set of conditions and strategies which proved universally applicable and successful. It is therefore also time to 'provincialize Europe'[92] once and for all, also in the field of disaster history. In

[91] Bankoff, 'Rendering the World Unsafe.' [92] Chakrabarty, *Provincializing Europe*.

every region, social vulnerability levels and societies' adaptability should be measured starting from the norms, institutions, technologies, and discourses prevailing in that particular region. Only then will it be possible to make a comparative assessment of different responses to similar pressures and hazards, without using a fixed standard or idea of success. Without retreating into relativism concerning social vulnerability and resilience towards hazards, this approach can distinguish between patterns and strategies in a non-Eurocentric way.

Our fourth suggestion is to analyze the unfolding of disasters from the local level. This may seem to go against the grain, now that global history figures prominently in historiography. This is not a plea to limit research to local actors and locally specific responses, however. The locality can and must be used as a starting point to look for broader patterns in dealing with hazards and disasters. Only at the local level do all types of responses, all the different actors, and all the different scales (the global, national, regional, and local level) come together. A top-down, centralized approach is often the most visible reaction, but the interaction between different actors operating at different levels is dominant through time and space. This is especially the case for social vulnerability. We can only grasp the impact of disasters by looking at how different social groups, or even the household or individual level, may have been affected. Local configurations of race, class, gender, age, profession, family, and household composition, etc., all had a tremendous impact on both the exposure to and the recovery from disaster. Moreover, as long as disaster history approaches disasters at the level of societies as a whole, it will almost invariably observe high levels of resilience: as we have seen, societal collapse and abrupt transformations are extremely rare in history. This societal continuity nevertheless masks big impacts from the level of social groups towards the individual onwards.[93] Similarly, it is important to move beyond large-scale or national responses, and look at how different levels interacted and affected responses to hazards. On a more practical level, existing localized histories and case studies, which can get lost in regional journals, should be brought to center stage by way of their intersection with overarching questions and bigger debates, so that comparative research becomes a possibility.

Finally, these pathways also require consideration and exploration of how (and, in many cases, whether) our results can inform not only disasters research, but also disaster management. It is clear that the value of the cautionary tale has its limits, but how do our results move beyond this? Does our historical laboratory offer us the basis to make

---

[93] Soens, 'Resilient Societies.'

predictive assessments about future disasters? Can our research intersect with, or feed into, agent-based modeling approaches that many researchers in the humanities and social sciences are wary of? These are likely to be contentious questions, but ones that an interdisciplinary field like historical disaster research cannot avoid. Whatever the answers, it is evident that, if historical disaster research is to speak to disaster management, and even policy, then a more transdisciplinary approach is called for – that is to say, research based upon co-design, co-production, and co-dissemination beyond the humanities and beyond the academic community as a whole.

7.5 A Final Word on Disaster Victims

Concluding a textbook on disaster history would not be possible without briefly returning to the single most essential question in the research field: why do people suffer from disaster, and how can such suffering be avoided or at least mitigated? It seems evident that disaster history should maintain a strong focus on the victims of disaster – the people struck by natural or human-made disaster, their misfortune and suffering, the causes of their suffering, and the efforts undertaken to alleviate or (even better) prevent this suffering. But, surprisingly, disaster victims are seldom at the heart of disaster history, in part because accurate data on casualties are often lacking, but also because historians have been more preoccupied by the causes of disaster, the coping mechanisms, or the recovery.[94] Historians are children of their time, and in the twenty-first-century 'Age of Disaster' there is a strong tendency to scrutinize history for examples of 'resilience': success and failure in adapting to hazards and disasters which – again – seem inevitable.

However, while adaptive processes are indeed very important in disaster management, they do not necessarily reduce the exposure to harm and suffering. History demonstrates that high levels of resilience can co-exist perfectly with high levels of vulnerability. For a number of cases – the Lisbon Earthquake of 1755, the 1693 Etna eruption on Sicily, and of course the fourteenth-century Black Death – historians have even argued that high numbers of casualties had a 'beneficial' impact on both the post-disaster recovery and the material welfare of the survivors, because of redistribution of wealth, investments in reconstruction, or improvements in policy making.[95] In resilience studies, this would be framed mainly as

---

[94] Even for the past century, data on people affected by or killed in a disaster are highly approximative: Eshghi & Larson, 'Disasters,' 72. See also Soens, 'Resilient Societies,' as well as Sections 2.2 and 6.1.1.
[95] Branca et al., 'Impacts,' 38.

an issue of scaling: the adaptive capacity of societies as a whole unfolds in different ways from the adaptive capacity of individual households.[96] In other words, what is bad for the individual can be good for the community. The expectation of high numbers of casualties gradually became part of a standard disaster narrative: a substantial death-count was expected – and was even needed in order to unleash relief and support from outside, especially when the disaster was situated in the colonial (or post-colonial) world and relief from the West had to be mobilized.[97]

Disaster history is still a very recent field of historical inquiry, gaining significant attention only from the late 1990s onwards. As a result, the field was not very much influenced by the older tradition of vulnerability studies, which aimed precisely to reveal the underlying, structural, causes of risk and harm.[98] Time and again, vulnerability studies demonstrated how exposure to risk and hazard was fueled by marginalization processes, which played themselves out both between regions and between households within a given region. Today, as well as in the past, risk is being 'dumped' on poor regions and people in marginal living conditions.[99] History has the unique potential to demonstrate how economic or political marginalization eventually led to vulnerability, but also to indicate ways in which this 'iron law' of vulnerability could be broken, by studying the conditions and instruments which allowed some societies to arrive at a more equal spread of vulnerability, or more 'inclusive' ways of achieving disaster resilience.

Perhaps one way forward for disaster history could be to shift attention from societies as a whole to communities and individual households and livelihoods, elaborating multidimensional assessments of household vulnerability and analyzing how households were capable of preventing disasters from happening and/or coped with their impact if they did happen. While disaster history has often questioned how regions recovered from disasters, historical studies investigating the livelihood trajectories of individual households struck by disaster remain rare. Moreover, in recent years, disaster studies have developed a clear interest in so-called 'traditional' coping mechanisms: strongly localized 'indigenous' knowledge on disaster prevention and mitigation, which was often passed from generation to generation, and might be 'reactivated' to complement 'modern' disaster prevention or relief mechanisms. Debates concentrate both on the relevance of highly localized coping mechanisms when confronted with increasingly globalized environmental, economic, and

---

[96] See Section 2.3.5.
[97] Bankoff, 'Rendering the World Unsafe'; see Smith, 'Volcanic Hazard in a Slave Society' for an early-nineteenth-century example of this logic.
[98] See Section 3.2.2. [99] Hillier & Castillo, *No Accident*, 4.

7.5 A Final Word on Disaster Victims

political pressures and on the intimate link between such traditional coping mechanisms and the empowerment of local populations, which might (or might not) be capable of organizing their environment in a way that fits their livelihood.[100] Historians can offer an important contribution to this debate, provided they reshuffle their analysis to focus on disaster victims, the causes of their vulnerability, and their capacity to impact disaster prevention, management, and recovery.

---

[100] Wong & Zhao, 'Living with Floods'; Hooli, 'Resilience of the Poorest.'

References

Acemoglu, Daron & Robinson, James A. 2013. *Why Nations Fail: The Origins of Power, Prosperity, and Poverty*. London: Profile Books.

Acemoglu, Daron, Robinson, James & Johnson, Simon. 2003. Disease and Development in Historical Perspective. *Journal of the European Economic Association* 1(2–3): 397–405.

Adamson, George & Nash, David. 2013. Long-Term Variability in the Date of Monsoon Onset over Western India. *Climate Dynamics* 40(11–12): 2589–2603.

Adamson, George & Nash, David. 2014. Documentary Reconstruction of Monsoon Rainfall Variability over Western India, 1781–1860. *Climate Dynamics* 42(3–4): 749–769.

Adamson, George, Rohland, Eleonora & Hannaford, Matthew. 2018. Re-thinking the Present: The Role of a Historical Focus in Climate Change Adaptation Research. *Global Environmental Change* 48(1): 195–205.

Adamson, George C. 2015. Private Diaries as Information Sources in Climate Research. *Wiley Interdisciplinary Reviews: Climate Change* 6(6): 599–611

Adger, W. Neil. 1999. Social Vulnerability to Climate Change and Extremes in Coastal Vietnam. *World Development* 27(2): 249–269.

Adger, W. Neil. 2010. Social Capital, Collective Action, and Adaptation to Climate Change, in Voss, Martin (ed.). *Der Klimawandel: Sozialwissenschaftliche Perspektiven*. Wiesbaden: Springer-Verlag, 327–345.

Adger, W. Neil & Brooks, Nick. 2003. Does Global Environmental Change Cause Vulnerability?, in Pelling, Mark (ed.). *Natural Disasters and Development in a Globalizing World*. London: Routledge, 19–42.

Adger, W. Neil, Brown, Katrina, Nelson, Donald R. et al. 2011. Resilience Implications of Policy Responses to Climate Change. *Wiley Interdisciplinary Reviews: Climate Change* 2(5): 757–766.

Adger, W. Neil, Dessai, Suraje, Goulden, Marisa et al. 2009. Are There Social Limits to Adaptation to Climate Change?, *Climatic Change* 93(3–4): 335–354.

Agarwal, Bina. 1992. Gender Relations and Food Security: Coping with Seasonality, Drought and famine in South Asia, in Beneria, L. & Feldman, S. (eds.). *Unequal Burden: Economic Crises, Persistent Poverty and Women's Work*. Boulder: Westview Press, 181–218.

Agarwal, Bina. 1995. *A Field of One's Own: Gender and Land Rights in South Asia*. Cambridge: Cambridge University Press.

Agrawal, Arun & Perrin, Nicolas. 2011. Climate Adaptation, Local Institutions and Rural Livelihoods, in Adger, William Neil, Lorenzoni, Irene &

O'Brien, Karen L. (eds.). *Adapting to Climate Change: Thresholds, Values, Governance*. Cambridge: Cambridge University Press, 350–367.

Aguirre, Benigo E. 2003. Better Disaster Statistics: The Lisbon Earthquake. *Journal of Interdisciplinary History* 43(1): 27–42.

Akasoy, Anna. 2006. Islamic Attitudes to Disasters in the Middle Ages: A Comparison of Earthquakes and Plagues. *The Medieval History Journal* 10 (1–2): 387–410.

Alam, Bhuiyan Monwar, Damole, Leticia N. & Wickramanayake, Ebel. 1996. Effects of Flood Mitigation Measure: Lessons from Dhaka Flood Protection Project, Bangladesh. *Asian Profile* 24(6): 511–524.

Alexander, David. 1997. The Study of Natural Disasters, 1977–97: Some Reflections on a Changing Field of Knowledge. *Disasters* 21(4): 284–304.

Alfani, Guido. 2010. The Effects of Plague on Distribution of Property: Ivrea, Northern Italy 1630. *Population Studies* 64(1): 61–75.

Alfani, Guido. 2011. The Famine of the 1590s in Northern Italy: An Analysis of the Greatest "System Shock" of Sixteenth Century. *Histoire et Mesure* 26(1): 17–50.

Alfani, Guido. 2013. *Calamities and the Economy in Renaissance Italy: The Grand Tour of the Horsemen of the Apocalypse*. Basingstoke: Palgrave MacMillan.

Alfani, Guido. 2013. Plague in Seventeenth-Century Europe and the Decline of Italy: An Epidemiological Hypothesis. *European Review of Economic History* 17(4): 408–430.

Alfani, Guido & Ammannati, Francesco. 2017. Long-Term Trends in Economic Inequality: The Case of the Florentine State, *ca.* 1300–1800. *Economic History Review* 70(4): 1072–1102.

Alfani, Guido & Di Tullio, Matteo. 2019. *The Lion's Share: Inequality and the Rise of the Fiscal State in Preindustrial Europe*. London & New York: Cambridge University Press.

Alfani, Guido & Murphy, Tommy E. 2017. Plague and Lethal Epidemics in the Pre-industrial World. *Journal of Economic History* 77(1): 314–343.

Alfani, Guido & Ó Gráda, Cormac (eds.). 2017. *Famine in European History*. Cambridge: Cambridge University Press.

Alfani, Guido & Ryckbosch, Wouter. 2015. Was There a "Little Convergence" in Inequality? Italy and the Low Countries Compared, *ca.* 1500–1800. *IGIER Working Papers*.

Ali, Imran. 1987. Malign Growth? Agricultural Colonization and the Roots of Backwardness in the Punjab. *Past & Present* 114(1): 110–132.

Allan, Rob, Endfield, Georgina, Damodaran, Vinita *et al.* 2016. Toward Integrated Historical Climate Research: The Example of Atmospheric Circulation Reconstructions over the Earth. *Wiley Interdisciplinary Reviews: Climate Change* 7(2): 164–174.

Álvarez-Nogal, Carlos & Prados de la Escosura, Leandro. 2013. The Rise and Fall of Spain (1270–1850). *Economic History Review* 66(1): 1–37.

Andersson, Lars Fredrik & Keskitalo, E. Carina H. 2016. Insurance Models and Climate Risk Assessments in a Historical Context. *Financial History Review* 23(2): 219–243.

Andreau, Jean. 1973. Histoire des séismes et histoire économique: Le tremblement de terre de Pompéi (62 ap. J.-C.). *Annales. Histoire, Sciences Sociales* 28(2): 369–395.
Appleby, Andrew B. 1979. Grain Prices and Subsistence Crises in England and France, 1590–1740. *Journal of Economic History* 39(4): 865–887.
Appuhn, Karl. 2010. Ecologies of Beef: Eighteenth Century Epizootics and the Environmental History of Early Modern Europe. *Environmental History* 15(2): 268–287.
Araújo, Ana Christina. 2006. The Lisbon Earthquake of 1755: Public Distress and Political Propaganda. *E-Journal of Portuguese History* 4(1): 1–25.
Arthi, Vellore. 2018. "The Dust Was Long in Settling": Human Capital and the Lasting Impact of the American Dust Bowl. *Journal of Economic History* 78(1): 196–230.
Ash, Eric H. 2004. *Power, Knowledge, and Expertise in Elizabethan England*. Baltimore: John Hopkins University Press.
Ash, Eric H. 2017. *The Draining of the Fens: Projectors, Popular Politics, and State Building*. Baltimore: Johns Hopkins University Press.
Atrash H. K. 2011. Parents' Death and Its Implications for Child Survival. *Revista brasileira de crescimento e desenvolvimento humano* 21(3): 759–770.
Baehrel, René. 1951. Epidémie et terreur: Histoire et sociologie. *Annales Historiques de la Révolution Française* 122: 113–146.
Baehrel, René. 1952. La haine de classe en temps d'épidémie. *Annales. Histoire, Sciences Sociales* 7(3): 351–360.
Baldwin, Peter. 2005. *Contagion and the State in Europe, 1830–1930*. Cambridge: Cambridge University Press.
Ballarini, M., Wallinga, J., Murray, A. S. et al. 2003. Optical Dating of Young Coastal Dunes on a Decadal Time Scale. *Quaternary Science Reviews* 22(10–13): 1011–1017.
Banik, Dan. 2007. Is Democracy the Answer? Famine Protection in Two Indian States, in Devereux, Stephen (ed.). *The New Famines: Why Famines Persist in an Era of Globalization*. London: Routledge, 290–311.
Bankoff, Greg. 2001. Rendering the World Unsafe: "Vulnerability" as Western Discourse. *Disasters* 25(1): 19–35.
Bankoff, Greg. 2003. *Cultures of Disaster: Society and Natural Hazards in the Philippines*. London: Routledge.
Bankoff, Greg. 2009. Cultures of Disaster, Cultures of Coping: Hazard as a Frequent Life Experience in the Philippines, in Mauch, Christof & Pfister, Christian (eds.). *Natural Disasters, Cultural Responses: Case Studies toward a Global Environmental History*. Lanham: Lexington Books, 265–284.
Bankoff, Greg. 2013. The "English Lowlands" and the North Sea Basin System: A History of Shared Risk. *Environment and History* 19(1): 3–37.
Bankoff, Greg. 2018. Remaking the World in Our Own Image: Vulnerability, Resilience and Adaptation as Historical Discourses. *Disasters* 43(2): 221–239.
Bankoff, Greg, Frerks, Georg & Hilhorst, Dorothea (eds.). 2004. *Mapping Vulnerability: Disasters, Development, and People*. London & Sterling: Routledge.
Barnes, Kent B. 2014. Social Vulnerability and Pneumonic Plague: Revisiting the 1994 Outbreak in Surat, India. *Environmental Hazards* 13(2): 161–180.

Barnes, Michele, Bodin, Örjan, Guerrero, Angela *et al.* 2017. The Social Structural Foundations of Adaptation and Transformation in Social-Ecological Systems. *Ecology and Society* 22(4): Article 16.

Barnett, Jon & O'Neill, Saffron. 2010. Maladaptation. *Global Environmental Change* 20(2): 211–213.

Barrett, Ron. 2008. The 1994 Plague in Western India: Human Ecology and the Risks of Misattribution, in Clunan, Anne L., Lavoy, Peter R. & Martin, Susan B. (eds.). *Terrorism, War or Disease? Unraveling the Use of Biological Weapons*. Stanford: Stanford University Press, 49–71.

Bauch, Martin & Schenk, Gerrit Jasper. 2019. *The Crisis of the 14th Century: Teleconnections between Environmental and Societal Change?* Boston: De Gruyter.

Bavel, Bas J. P. van. 2002. People and Land: Rural Population Developments and Property Structures in the Low Countries, c. 1300–c. 1600. *Continuity and Change* 17(1): 9–37.

Bavel, Bas J. P. van. 2010. *Manors and Markets: Economy and Society in the Low Countries 500–1600*. Oxford: Oxford University Press.

Bavel, Bas J. P. van. 2016. *The Invisible Hand?: How Market Economies Have Emerged and Declined since AD 500*. Oxford: Oxford University Press.

Bavel, Bas J. P. van & Curtis, Daniel R. 2016. Better Understanding Disasters by Better Using History: Systematically Using the Historical Record as One Way to Advance Research into Disasters. *International Journal of Mass Emergencies and Disasters* 34(1): 143–169.

Bavel, Bas J. P. van, Curtis, Daniel R., Hannaford, Matthew *et al.* 2019. Climate and Society in Long-Term Perspective: Opportunities and Pitfalls in the Use of Historical Datasets. *Wiley Interdisciplinary Reviews: Climate Change* 10(6): Article e611.

Bavel, Bas J. P. van, Curtis, Daniel R. & Soens, Tim. 2018. Economic Inequality and Institutional Adaptation in Response to Flood Hazards: A Historical Analysis. *Ecology and Society* 23(4): Article 30.

Bavel, Bas J. P. van & Rijpma, Auke. 2016. How Important Were Formalized Charity and Social Spending before the Rise of the Welfare State? A Long-Run Analysis of Selected Western European Cases, 1400–1850. *Economic History Review* 69(1): 159–187.

Bavel, Bas J. P. van & Zanden, Jan Luiten van. 2004. The Jump-Start of the Holland Economy during the Late-Medieval Crisis, c. 1350–c. 1500. *Economic History Review* 57(3): 503–532.

Baydal Sala, Vicent & Esquilache Martí, Ferran. 2014. Exploitation and Differentiation: Economic and Social Stratification in the Rural Muslim Communities of the Kingdom of Valencia, 13th–16th centuries, in Aparisi, Frederic & Royo, Vicent (eds.). *Beyond Lords and Peasants. Rural Elites and Economic Differentiation in Pre-modern Europe*. Valencia: Publicacions de la Universitat de València, 37–67.

Beck, Tony & Nesmith, Cathy. 2001. Building on Poor People's Capacities: The Case of Common Property Resources in India and West Africa. *World Development* 29(1): 119–133.

Beck, Ulrich, Lash, Scott & Wynne, Brian. 1992. *Risk Society: Towards a New Modernity*. London: SAGE Publications.

Beckert, Sven. 2015. *Empire of Cotton: A Global History*. New York: Vintage.
Behling, Noriko, Williams, Mark C., Behling, Thomas G. & Managi, Shunsuke. 2019. Aftermath of Fukushima: Avoiding Another Nuclear Disaster. *Energy Policy* 126: 411–420.
Behringer, Wolfgang. 1999. Climatic Change and Witch-Hunting: The Impact of the Little Ice Age on Mentalities. *Climatic Change* 43(1): 335–351.
Bell, Clive & Lewis, Maureen. 2005. Economic Implications of Epidemics Old and New. *Center for Global Development Working Paper* 54.
Beltrán Tapia, Francisco J. 2015. Social and Environmental Filters to Market Incentives: The Persistence of Common Land in Nineteenth-Century Spain. *Journal of Agrarian Change* 15(2): 239–260.
Béné, Christophe, Al-Hassan, Ramatu M., Amarasinghe, Oscar et al. 2016. Is Resilience Socially Constructed? Empirical Evidence from Fiji, Ghana, Sri Lanka, and Vietnam. *Global Environmental Change* 38(1): 153–170.
Benedictow, Ole. 2012. New Perspectives in Medieval Demography: The Medieval Demographic System, in Bailey, Mark & Rigby, Stephen (eds.). *Town and Countryside in the Age of the Black Death: Essays in Honour of John Hatcher*. Turnhout: Brepols, 3–42.
Benedictow, Ole J. 2016. *The Black Death and Later Plague Epidemics in the Scandinavian Countries: Perspectives and Controversies*. Berlin: De Gruyter.
Berkes, Fikret & Folke, Carl (eds.). 1998. *Linking Social and Ecological Systems: Management Practices and Social Mechanisms for Building Resilience*. Cambridge: Cambridge University Press.
Berland, Alexander J. & Endfield, Georgina. 2018. Drought and Disaster in a Revolutionary Age: Colonial Antigua during the American Independence War. *Environment and History* 24(2): 209–235.
Berren, Michael R., Beigel, Allan & Barker, Gypsy. 1982. A Typology for the Classification of Disasters: Implications for Intervention. *Community Mental Health Journal* 18(2): 120–134.
Bever, Edward. 2002. Witchcraft, Female Aggression, and Power in the Early Modern Community. *Journal of Social History* 35(4): 955–988.
Beveridge, William H. 1922. Wheat Prices and Rainfall in Western Europe. *Journal of the Royal Statistical Society* 85(3): 412–475.
Biraben, Jean-Noël. 1975/1976. *Les hommes et la peste en France et dans les pays européens et méditerranéens*, 2 volumes. Paris: Mouton.
Blaikie, Piers, Cannon, Terry, Davis, Ian & Wisner, Ben. 1994. *At Risk: Natural Hazards, People's Vulnerability, and Disasters*. London: Routledge.
Blair, Robert A., Morse, Benjamin S. & Tsai, Lily L. 2017. Public Health and Public Trust: Survey Evidence from the Ebola Virus Disease Epidemic in Liberia. *Social Science & Medicine* 172: 89–97.
Blažina Tomić, Zlata & Blažina, Vesna. *Expelling the Plague: The Health Office and the Implementation of Quarantine in Dubrovnik, 1377–1533*. Montreal: McGill-Queens University Press.
Bloom, David E., Canning, David & Fink, Günther. 2014. Disease and Development Revisited. *Journal of Political Economy* 122(6): 1355–1366.

Bodin, Örjan. 2017. Collaborative Environmental Governance: Achieving Collective Action in Social-Ecological systems. *Science* 6352(357): Article eaan1114.
Bolòs, Jordi. 2001. Changes and Survival: The Territory of Lleida (Catalonia) After the Twelfth-Century Conquest. *Journal of Medieval History* 27(4): 313–329.
Bolton, J. L. 2013. Looking for *Yersinia pestis*: Scientists, Historians and the Black Death, in Clark, Linda & Rawcliffe, Carole (eds.). *The Fifteenth Century XII: Society in an Age of the Plague*. Woodbridge: The Boydell Press, 15–38.
Borsch, Stuart J. 2004. Environment and Population: The Collapse of Large Irrigation Systems Reconsidered. *Comparative Studies in Society and History* 46 (3): 451–468.
Borsch, Stuart J. 2005. *The Black Death in Egypt and England: A Comparative Study*. Austin: University of Texas Press.
Borsch, Stuart J. 2014. Plague Depopulation and Irrigation Decay in Medieval Egypt, in Green, Monica H. (ed.). *Pandemic Disease in the Medieval World: Rethinking the Black Death. The Medieval Globe*, Volume 1. Kalamazoo & Bradford: Arc Medieval Press, 125–156.
Borsch, Stuart J. & Sabraa, Tarek. 2017. Refugees of the Black Death: Quantifying Rural Migration for Plague and Other Environmental Disasters. *Annales de Démographie Historique* 134(2): 63–93.
Bowler, Peter J. & Rhys Morus, Iwan. 2005. *Making Modern Science: A Historical Survey*. Chicago & London: The University of Chicago Press.
Bowman Jannetta, Ann. 1987. *Epidemics and Mortality in Early Modern Japan*. Princeton: Princeton University Press.
Bowsky, William M. 1964. The Impact of the Black Death upon Sienese Government and Society. *Speculum* 39(1): 1–34.
Branca, Stefano, Azzaro, Raffaele, De Beni, Emanuela, Chester, David & Duncan, Angus. 2015. Impacts of the 1669 Eruption and the 1693 Earthquakes on the Etna Region (Eastern Sicily, Italy): An Example of Recovery and Response of a Small Area to Extreme Events. *Journal of Volcanology and Geothermal Research* 303(1): 25–40.
Brantz, Dorothee. 2011. Risky Business: Disease, Disaster and the Unintended Consequences of Epizootics in Eighteenth- and Nineteenth-Century France and Germany. *Environment and History* 17(1): 35–51.
Braun, Joachim von. 1991. *A Policy Agenda for Famine Prevention in Africa*. Washington D.C.: International Food Policy Research Institute.
Brecke, Peter. 1999. Violent Conflicts 1400 A.D. to the Present in Different Regions of the World. *Meeting of the Peace Science Society (International)*. Ann Arbor, Michigan, 8–10 October.
Brecke, Peter. 2012. Notes Regarding the Conflict Catalog. *Center for Global Economic History*.
Brewis, Georgina. 2010. Fill Full the Mouth of Famine: Voluntary Action in Famine Relief in India, 1896–1901. *Modern Asian Studies* 44(4): 887–918.
Briggs, Asa. 1961. Cholera and Society in the Nineteenth Century. *Past & Present* 19(1): 76–96.

Briggs, Robin. 1996. "Many Reasons Why": Witchcraft and the Problem of Multiple Explanation, in Barry, Jonathan, Hester, Marianne & Roberts, Gareth (eds.). *Witchcraft in Early Modern Europe: Studies in Culture and Belief*. Cambridge: Cambridge University Press, 49–63.

Britnell, Richard H. 1996. *The Commercialisation of English Society, 1000–1500*. Manchester: Manchester University Press.

Broadberry, Stephen, Campbell, Bruce M. S., Klein, Alexander, Overton, Mark & Leeuwen, Bas van. 2015. *British Economic Growth, 1270–1870*. Cambridge: Cambridge University Press.

Brook, Timothy. 2010. *The Troubled Empire: China in the Yuan and Ming Dynasties*. Cambridge: Harvard University Press.

Brook, Timothy. 2017. Differential Effects of Global and Local Climate Data in Assessing Environmental Drivers of Epidemic Outbreaks. *Proceedings of the National Academy of Sciences* 114(49): 12845–12847.

Büntgen, Ulf, Ginzler, Christian, Esper, Jan, Tegel, Willy & McMichael, Anthony J. 2012. Digitizing Historical Plague. *Clinical Infectious Diseases* 55(11): 1586–1588.

Büntgen, Ulf, Myglan, Vladimir S., Charpentier Ljungqvist, Fredrik et al. 2016. Cooling and Societal Change during the Late Antique Little Ice Age from 536 to Around 660 AD. *Nature Geoscience* (9): 231–236.

Büntgen, Ulf, Tegel, Willy, Nicolussi, Kurt et al. 2011. 2500 Years of European Climate Variability and Human Susceptibility. *Science* 331 (6017): 578–582.

Burchi, Francesco. 2011. Democracy, Institutions and Famines in Developing and Emerging Countries. *Canadian Journal of Development Studies* 32(1): 17–31.

Butzer, Karl W. 2012. Collapse, Environment, and Society. *Proceedings of the National Academy of Sciences* 109(10): 3632–3639.

Butzer, Karl W. & Endfield, Georgina H. 2012. Critical Perspectives on Historical Collapse. *Proceedings of the National Academy of Sciences* 109(10): 3628–3631.

Calain, Philippe & Poncin, Marc. 2017. Reaching out to Ebola Victims: Coercion, Persuasion or an Appeal for Self-Sacrifice?, in Messelken, Daniel & Winkler, David (eds.). *Ethical Challenges for Military Health Care Personnel: Dealing with Epidemics*. London: Routledge, Chapter 6.

Camenisch, Chantal & Rohr, Christian. 2018. When the Weather Turned Bad: The Research of Climate Impacts on Society and Economy during the Little Ice Age in Europe. An Overview. *Geographical Research Letters* 44(1): 99–114.

Campbell, Bruce M. S. 2005. The Agrarian Problem in the Early 14th Century. *Past & Present* 188(1): 3–70.

Campbell, Bruce M. S. 2010. Nature as Historical Protagonist: Environment and Society in Pre-industrial England. *Economic History Review* 63(2): 281–314.

Campbell, Bruce M. S. 2016. *The Great Transition: Climate, Disease and Society in the Late-Medieval World*. Cambridge: Cambridge University Press.

Campopiano, Michele. 2011. Rural Communities, Land Clearance and Water Management in the Po Valley in the Central and Late Middle Ages. *Journal of Medieval History* 39(4): 377–393.

Campopiano, Michele. 2013. The Evolution of the Landscape and the Social and Political Organisation of Water Management: The Po Valley in the Middle Ages (Fifth to Fourteenth Centuries), in Thoen, Eric, Borger, Guus J., Soens, Tim et al. (eds.). *Landscapes or Seascapes? The History of the Coastal Environment in the North Sea Area Reconsidered.* CORN Publication Series, Volume 13. Turnhout: Brepols, 313–332.

Cannon, Terry & Müller-Mahn, Detlef. 2010. Vulnerability, Resilience and Development Discourses in Context of Climate Change. *Natural Hazards* 55 (3): 621–635.

Cao, Shuji & Li, Yushang. 2006. *Shuyi, zhanzhengyu heping: Zhongguo de huanjingyu shehui bianqian, 1230–1960* [Plague, War and Peace: Environment and Social Change in China, 1230–1960]. Ji'nan: Shandong huabao chubanshe.

Carmichael, Ann G. 1986. *Plague and the Poor in Renaissance Florence.* Cambridge: Cambridge University Press.

Carmichael, Ann G. 1998. The Last Past Plague: The Uses of Memory in Renaissance Epidemics. *Journal of the History of Medicine and Allied Sciences* 53(2): 132–160.

Carmichael, Ann G. 2008. Universal and Particular: The Language of Plague, 1348–1500. *Medical History* 52(S27): 17–52.

Carmichael, Sarah G., Pleijt, Alexandra de, Zanden, Jan Luiten van & Moor, Tine de. 2016. The European Marriage Pattern and Its Measurement. *Journal of Economic History* 76(1): 196–204.

Carpenter, Stephen. R. & Brock, William A. 2008. Adaptive Capacity and Traps. *Ecology and Society* 13 (2): Article 40.

Carré, Matthieu, Azzoug, Moufok, Zaharias, Paul et al. 2018. Modern Drought Conditions in Western Sahel Unprecedented in the Past 1600 Years. *Climate Dynamics* 52(3–4): 1949–1964.

Carter, Michael. R., Little, Peter D., Mogues, Tewodaj & Negatu, Workneh. 2007. Poverty Traps and Natural Disasters in Ethiopia and Honduras. *World Development* 35(5): 835–856.

Casari, Marco. 2007. Emergence of Endogenous Legal Institutions: Property Rights and Community Governance in the Italian Alps. *Journal of Economic History* 67(1): 191–226.

Casson, Mark & Casson, Catharine. 2015. Economic Crises in England, 1270–1520: A Statistical Approach, in Brown, A. T., Burn, Andy & Doherty, Rob (eds.). *Crises in Economic and Social History: A Comparative Perspective.* Woodbridge: The Boydell Press, 79–110.

Castel, Ilona I. Y. 1991. *Late Holocene Eolian Drift Sands in Drenthe (The Netherlands).* PhD Thesis: Utrecht University.

Cavallo, Eduardo, Galiani, Sebastian, Noy, Ilan & Pantano, Juan. 2013. Catastrophic Natural Disasters and Economic Growth. *Review of Economics and Statistics* 95(5): 1549–1561.

Chakrabarty, Dipesh. 2000. *Provincializing Europe: Postcolonial Thought and Historical Difference.* Princeton: Princeton University Press.

Chakrabarty, Dipesh. 2009. The Climate of History: Four Theses. *Critical Inquiry* 35(2): 197–222.

Chambers, Robert. 1989. Vulnerability, Coping and Policy. *IDS Bulletin* 20 (2): 1–7.
Che, Vivian B., Kervyn, Matthieu, Ernst, G. G. J. et al. 2011. Systematic Documentation of Landslide Events in Limbe Area (Mt. Cameroon Volcano, SW Cameroon): Geometry, Controlling and Triggering Factors. *Natural Hazards* 59(1): 47–74.
Chen, Yuyu & Zhou, Li-An. 2007. The Long-Term Health and Economic Consequences of the 1959–1961 Famine in China. *Journal of Health Economics* 26(4): 659–681.
Christakos, George, Olea, Ricardo A., Serre, Marc L., Yu, Hwa-Lung & Wang, Lin-Lin. 2005. *Interdisciplinary Public Health Reasoning and Epidemic Modelling: The Case of Black Death*. Berlin: Springer.
Cipolla, Carlo M. 1972. *The Economic History of World Population*. London: Penguin Publishers.
Cipolla, Carlo M. 1974. The Plague and the Pre-Malthus Malthusians. *Journal of European Economic History* 3(2): 277–284.
Clark, Brett & Foster, John B. 2006. The Environmental Conditions of the Working Class: An Introduction to Selections from Frederick Engels's *The Condition of the Working Class in England in 1844*. *Organization & Environment* 19(3): 375–388.
Coetzee, Christo & Niekerk, Dewald van. 2012. Tracking the Evolution of the Disaster Management Cycle: A General System Theory Approach. *Jàmbá: Journal of Disaster Risk Studies* 4(1): 1–9.
Cohen, Charles & Werker, Eric D. 2008. The Political Economy of "Natural" Disasters. *Journal of Conflict Resolution* 52(6): 795–819.
Cohn Jr., Samuel K. 2017. Patterns of Plague in Late Medieval and Early-Modern Europe, in Jackson, Mark (ed.). *The Routledge History of Disease*. London & New York: Routledge, 165–182.
Cohn Jr., Samuel K. 2002. *The Black Death Transformed: Disease and Culture in Early Renaissance Europe*. London & New York: Arnold & Oxford University Press.
Cohn Jr., Samuel K. 2007. The Black Death and the Burning of Jews. *Past & Present* 196(1): 3–36.
Cohn Jr., Samuel K. 2008. *Lust for Liberty: The Politics of Social Revolt in Medieval Europe, 1200–1425: Italy, France, and Flanders*. Cambridge: Harvard University Press.
Cohn Jr., Samuel K. 2010. *Cultures of Plague: Medical Thinking at the End of the Renaissance*. Oxford: Oxford University Press.
Cohn Jr., Samuel K. 2016. Rich and Poor in Western Europe, c. 1375–1475: The Political Paradox of Material Well-Being, in Farmer, Sharon A. (ed.). *Approaches to Poverty in Medieval Europe: Complexities, Contradictions, Transformations, c. 1100–1500*. Turnhout: Brepols, 145–173.
Cohn Jr., Samuel K. 2017. Cholera Revolts: A Class Struggle We May Not Like. *Social History* 42(2): 162–180.
Cohn Jr., Samuel K. 2018. *Epidemics: Hate and Compassion from the Plague of Athens to AIDS*. Oxford: Oxford University Press.

Cohn Jr., Samuel K. & Kutalek, Ruth. 2016. Historical Parallels, Ebola Virus Disease and Cholera: Understanding Community Distrust and Social Violence with Epidemics. *PLOS Currents Outbreaks* 8(1): Article PMC4739438.

Colet, Anna, Muntané i Santiveri, Xavier J., Ruíz Ventura, Jordi *et al.* 2016. The Black Death and Its Consequences for the Jewish Community in Tàrrega: Lessons from History and Archeology, in Green, Monica H. (ed.). *Pandemic Disease in the Medieval World: Rethinking the Black Death. The Medieval Globe*, Volume 1. Kalamazoo & Bradford: Arc Medieval Press, 63–96.

Collet, Dominik & Schuh, Maximilian (eds). 2018. *Famines during the "Little Ice Age" (1300–1800)*. Cham: Springer.

Colton, Craig E. 2002. Basin Street Blues: Drainage and Environmental Equity in New Orleans, 1890–1930. *Journal of Historical Geography* 28(2): 237–257.

Condorelli, Stéfano. 2006. The Reconstruction of Catania after the Earthquake of 1693, in Dunkeld, Malcolm, Campbell, James W. P., Louw, Hentie *et al.* (eds.). *Proceedings of the Second International Congress on Construction History*, Volume 1. Exeter: Short Run Press, 799–816.

Condorelli, Stéfano. 2011. *"U tirrimotu ranni": Lectures du tremblement de terre de Sicile de 1693*. Unpublished Thesis, Geneva: Université de Genève.

Congost, Rosa. 2003. Property Rights and Historical Analysis: What Rights? What History? *Past & Present* 181(1): 73–106.

Coomans, J. 2018. *In Pursuit of a Healthy City: Sanitation and the Common Good in the Late Medieval Low Countries*. PhD Thesis: Amsterdam School of Historical Studies.

Costanza, Robert, Graumlich, Lisa J. & Steffen, Will (eds.). 2007. *Sustainability or Collapse? An Integrated History and Future of People on Earth*. Cambridge: MIT Press.

Costanza, Robert, Graumlich, Lisa, Steffen, Will *et al.* 2007. Sustainability or Collapse: What Can We Learn from Integrating the History of Humans and the Rest of Nature? *AMBIO: A Journal of the Human Environment* 36(7): 522–527.

Covello, Vincent T., Winterfeldt, Detlof von & Slovic, Paul. 1988. Risk Communication, in Travis, Curtis C. (ed.). *Carcinogen Risk Assessment: Contemporary Issues in Risk Analysis*, Volume 3. Boston: Springer, 193–207.

Crona, Beatrice & Bodin, Örjan. 2010. Power Asymmetries in Small-Scale Fisheries: A Barrier to Governance Transformability? *Ecology and Society* 15(4): Article 32.

Crosby, Alfred W. 2009. *The Measure of Reality: Quantification and Western Society, 1250–1600*. Cambridge: Cambridge University Press.

Crutzen, Paul J. 2002. Geology of Mankind: The Anthropocene. *Nature* 415: 23.

Cruyningen, Piet van. 2014. From Disaster to Sustainability: Floods, Changing Property Relations and Water Management in the South-Western Netherlands, c. 1500–1800. *Continuity and Change* 29(2): 241–265.

Cui, Yujun, Yu, Chang, Yan, Yanfeng *et al.* 2013. Historical Variations in Mutation Rate in an Epidemic Pathogen, *Yersinia pestis*. *Proceedings of the National Academy of Sciences* 110(2): 577–582.

Cumming, Graeme S. & Collier, John. 2005. Change and Identity in Complex Systems. *Ecology and Society* 10(1): Article 29.

Cumming, Graeme S., Cumming, David H. M. & Redman, Charles L. 2006. Scale Mismatches in Social-Ecological Systems: Causes, Consequences, and Solutions. *Ecology and Society* 11(1): Article 14.

Curtis, Daniel R. 2013. Is There an Agro-town Model for Southern Italy? Exploring the Diverse Roots and Development of the Agro-town Structure through a Comparative Case Study in Apulia. *Continuity & Change* 28(3): 377–419.

Curtis, Daniel R. 2014. *Coping with Crisis: The Resilience and Vulnerability of Pre-industrial Settlements*. Farnham: Ashgate.

Curtis, Daniel R. 2014. The Impact of Land Accumulation and Consolidation on Population Trends in the Pre-industrial Period: Two Contrasting Cases in the Low Countries. *Historical Research* 87(236): 194–228.

Curtis, Daniel R. 2015. An Agro-town Bias? Re-examining the Micro-demographic Model for Southern Italy in the Eighteenth Century. *Journal of Social History* 48(3): 685–713.

Curtis, Daniel R. 2016. Danger and Displacement in the Dollard: The 1509 Flooding of the Dollard Sea (Groningen) and Its Impact on Long-Term Inequality in the Distribution of Property. *Environment and History*, 22(1): 103–135.

Curtis, Daniel R. 2016. Did the Commons Make Medieval and Early Modern Rural Societies More Equitable? A Survey of Evidence from across Western Europe, 1300–1800. *Journal of Agrarian Change* 16(4): 646–664.

Curtis, Daniel R. 2016. The Emergence of Concentrated Settlements in Medieval Western Europe: Explanatory Frameworks in the Historiography. *Canadian Journal of History* 48(2): 223–251.

Curtis, Daniel. R. 2016. Was Plague an Exclusively Urban Phenomenon? Plague Mortality in the Seventeenth-Century Low Countries. *Journal of Interdisciplinary History* 47(2): 139–170.

Curtis, Daniel R. 2020. All Equal in the Presence of Death? Epidemics and Redistribution in the Pre-industrial Period, in Thoen, Erik & Alfani, Guido (eds.). *Inequality and Development in Rural Europe (Middle Ages–19th century)*. CORN Publication Series. Turnhout: Brepols.

Curtis, Daniel R. 2020. Preserving the Ordinary: Social Resistance during Second Pandemic Plagues in the Low Countries, in Gerrard, Chris, Brown, Peter & Forlin, Paolo (eds.). *Waiting for the Ends of the World: Perceptions of Disaster and Risk in Medieval Europe*. Society for Medieval Archaeology, Volume 43. London: Routledge.

Curtis, Daniel R., Bavel, Bas J. P. van. & Soens, Tim. 2016. History and the Social Sciences: Shock Therapy with Medieval Economic History as the Patient. *Social Science History* 40(4): 751–774.

Curtis, Daniel R. & Campopiano, Michele. 2014. Medieval Land Reclamation and the Creation of New Societies: Comparing Holland and the Po Valley, *c.* 800–*c.* 1500. *Journal of Historical Geography* 44(1): 93–108.

Curtis, Daniel R. & Dijkman, Jessica. 2018. The Escape from Famine in the Northern Netherlands: A Reconsideration Using the 1690s Harvest Failures and a Broader Northwest European Perspective. *The Seventeenth Century* 34(2): 229–258.

Curtis, Daniel R. & Han, Qijun. 2020. The Female Mortality Advantage in the Seventeenth-Century Rural Low Countries. *Gender and History* (forthcoming).

Curtis, Daniel R. & Roosen, Joris. 2018. On the Importance of History. *Inference: International Review of Science* 4(2), https://inference-review.com/letter/on-the-importance-of-history.

D'Souza, Rohan. 2006. *Drowned and Dammed: Colonial Capitalism and Flood Control in Eastern India.* New Delhi: Oxford University Press.

Da Molin, Giovanna. 1990. Family Forms and Domestic Service in Southern Italy from the Seventeenth to the Nineteenth Centuries. *Journal of Family History* 15(4): 503–527.

Dalfes, H. Nüzhet, Kukla, George & Weiss, Harvey (eds.). 1997. *Third Millennium BC Climate Change and Old World Collapse.* Heidelberg: Springer.

Dam, Petra J. E. M. van 2012. Denken over natuurrampen: Overstromingen en de amfibische cultuur. *Tijdschrift voor Waterstaatsgeschiedenis* 21(1): 1–10.

Dam, Petra J. E. M. van. 2017. An Amphibious Culture: Coping with Floods in the Netherlands, in Coates, Peter, Moon, David & Warde, Paul (eds.). *Local Places, Global Processes: Histories of Environmental Change in Britain and Beyond.* Oxford: Oxford Books, 78–93.

Daniell, James E., Wenzel, Friedemann & Khazai, Bijan. 2010. The Cost of Historic Earthquakes Today – Economic Analysis since 1900 through the use of CATDAT. AEES 2010 Conference, Perth, Australia 21: Paper No. 7.

Das Gupta, Monica & Shuzhuo, Li. 1999. Gender Bias in China, South Korea and India, 1920–1990: The Effects of War, Famine, and Fertility Decline. *Development and Change* 30(3): 619–652.

Davis, Mike. 2002. *Late Victorian Holocausts: El Niño Famines and the Making of the Third World.* London: Verso.

Dean, Trevor. 2015. Plague and Crime: Bologna, 1348–1351. *Continuity and Change* 30(3): 367–393.

Dearing, John A., Wang, Rong, Zhang, Ke et al. 2014. Safe and Just Operating Spaces for Regional Social-Ecological Systems. *Global Environmental Change* 28(1): 227–238.

Degroot, Dagomar. 2018. *The Frigid Golden Age: Climate Change, the Little Ice Age, and the Dutch Republic, 1560–1720.* Cambridge: Cambridge University Press.

Delaney, Alyne Elizabeth. Taking the High Ground: The Impact of Public Policy on Rebuilding Neighborhoods in Coastal Japan after the 2011 Great East Japan Earthquake and Tsunami, in Companion, Michèle (ed.). *Disaster's Impact on Livelihood and Cultural Survival: Losses, Opportunities and Mitigation.* Boca Raton: CRC Press, 63–74.

Delumeau, Jean. 1978. *La peur en Occident: XIVe–XVIIIe siècles.* Paris: Fayard.

Dennison, Tracy & Ogilvie, Sheilagh. 2014. Does the European Marriage Pattern Explain Economic Growth? *Journal of Economic History* 74(3): 651–693.

Dennison, Tracy K. & Ogilvie, Sheilagh. 2007. Serfdom and Social Capital in Bohemia and Russia. *Economic History Review* 60(3): 513–544.

Derese, Cilia, Vandenberghe, Dimity, Eggermont, Nele et al. 2010. A Medieval Settlement Caught in the Sand: Optical Dating of Sand-Drifting at Pulle (N Belgium). *Quaternary Geochronology* 5(2–3): 336–341.

Devereux, Stephen. 1993. Goats before Ploughs: Dilemmas of Household Response Sequencing during Food Shortages. *IDS Bulletin* 24(4): 52–59.

Devereux, Stephen, Sida, Lewis & Nelis, Tina. 2017. *Famine: Lessons Learned*. Brighton: IDS.

Devereux, Stephen. 2002. The Malawi Famine of 2002. *IDS Bulletin* 33(4): 70–78.

Devereux, Stephen (ed.). 2007. *The New Famines: Why Famines Persist in an Age of Globalization*. London & New York: Routledge.

Devereux, Stephen. 2009. Why Does Famine Persist in Africa? *Food Security* 1(1): 25–35.

DeWitte, Sharon N. 2014. Health in Post-Black Death London (1350–1538): Age Patterns of Periosteal New Bone Formation in a Post-epidemic Population. *American Journal of Physical Anthropology* 155(2): 260–267

DeWitte, Sharon N. 2016. The Anthropology of Plague: Insights from Bioarcheological Analyses of Epidemic Cemeteries, in Green, Monica H. (ed.). *Pandemic Disease in the Medieval World: Rethinking the Black Death. The Medieval Globe*, Volume 1. Kalamazoo & Bradford: Arc Medieval Press, 97–123.

DeWitte, Sharon N. 2018. Stress, Sex, and Plague: Patterns of Developmental Stress and Survival in Pre- and Post-Black Death London. *American Journal of Human Biology* 30: Article e23073.

Di Tullio, Matteo. 2014. *The Wealth of Communities: War, Resources and Cooperation in Renaissance Lombardy*. Farnham: Ashgate.

Di Tullio, Matteo. 2017. Cooperating in Time of Crisis: War, Commons, and Inequality in Renaissance Lombardy. *Economic History Review* 71(1): 82–105.

Diamond, Jared. 2005. *Collapse: How Societies Choose to Fail or Succeed*. New York: Penguin Group.

Dietz, T., Stern, P. C., & Rycroft, R. W. 1989. Definitions of Conflict and the Legitimation of Resources: The Case of Environmental Risk. *Sociological Forum* 4(1): 47–70.

Dijkman, Jessica. 2017. Coping with Scarcity: A Comparison of Dearth Policies of Three Regions in Northwestern Europe in the Fifteenth and Sixteenth Centuries. *Tijdschrift voor Economische en Sociale Geschiedenis* 14(3): 5–30.

Dijkman, Jessica. 2018. Bread for the Poor: Poor Relief and the Mitigation of the Food Crises of the 1590s and the 1690s in Berkel, Holland, in Collet, Dominik & Schuh, Maximilian (eds.). *Famines during the "Little Ice Age" (1300–1800)*. Cham: Springer, 171–193.

Dijkman, Jessica. 2019. Feeding the Hungry: Poor Relief and Famine in Northwestern Europe, 1500–1700, in Dijkman, Jessica & Leeuwen, Bas van (eds.). *An Economic History of Famine Resilience*. London: Routledge, 93–111.

Dincecco, Mark & Onorato, Massimiliano G. 2016. Military Conflict and the Rise of Urban Europe. *Journal of Economic Growth* 21(3): 259–282.

Dobson, Mary J. 1992. Contours of Death: Disease, Mortality and the Environment in Early Modern England. *Health Transition Review* 2: 77–95.

Dobson, Mary J. 1997 *Contours of Death and Disease in Early Modern England*. Cambridge: Cambridge University Press.

Dovers, Stephen R. & Hezri, Adnan A. 2010. Institutions and Policy Processes: The Means to the Ends of Adaptation. *Wiley Interdisciplinary Reviews: Climate Change* 1(2): 212–231.

Duara, Prasenjit. 1990. Ten Elites and the Structures of Authority in the Villages of North China, 1900–1949, in Esherick, Joseph W. & Rankin, Mary B. (eds.). *Chinese Local Elites and Patterns of Dominance.* Berkeley: University of California Press, 262–282.

Duffield, John S. 2016. Japanese Energy Policy after Fukushima Daiichi: Nuclear Ambivalence. *Political Science Quarterly* 131(1): 133–162.

Dugmore, Andrew J., McGovern, Thomas H., Vésteinsson, Orri et al. 2012. Cultural Adaptation, Compounding Vulnerabilities and Conjunctures in Norse Greenland. *Proceedings of the National Academy of Sciences* 109(10): 3658–3663.

Dunstan, Helen. 1975. The Late Ming Epidemics: A Preliminary Survey. *Ch'ing-shih wen-t'i* 3(3): 1–59.

Dyer, Christopher. 1989. *Standards of Living in the Later Middle Ages: Social Change in England c. 1200–1520.* Cambridge: Cambridge University Press.

Dyer, Christopher. 1996. Taxation and Communities in Late Medieval England, in Britnell, Richard H. & Hatcher, John (eds.). *Progress and Problems in Medieval England: Essays in Honour of Edward Miller.* Cambridge: Cambridge University Press.

Dyer, Christopher. 2007. *An Age of Transition? Economy and Society in England in the Later Middle Ages.* Oxford: Oxford University Press.

Dyson, Tim & Ó Gráda, Cormac. 2002. Introduction, in Dyson, Tim & Ó Gráda, Cormac (eds.). *Famine Demography: Perspectives from the Past and Present.* Oxford: Oxford University Press, 1–19.

Ebert, John David. 2012. *The Age of Catastrophe: Disaster and Humanity in Modern Times.* Jefferson & London: McFarland.

Edgerton-Tarpley, Katheryn. 2004. Family and Gender in Famine: Cultural Responses to Disaster in North China, 1876–1879. *Journal of Women's History* 16(4): 119–147.

Edgerton-Tarpley, Katheryn. 2008. *Tears from Iron: Cultural Responses to Famine in Nineteenth-Century China.* Berkeley, Los Angeles & London: University of California Press.

Eichelberger, Laura. 2007. SARS and New York's Chinatown: The Politics of Risk and Blame during an Epidemic of Fear. *Social Science & Medicine* 65(6): 1284–1295.

Elliott, James R. & Pais, Jeremy. 2006. Race, Class, and Hurricane Katrina: Social Differences in Human Responses to Disaster. *Social Science Research* 35(2): 295–321.

Ellis, Erle C. 2018. *Anthropocene: A Very Short Introduction.* Oxford & New York: Oxford University Press.

Elvin, Mark. 1993. Three Thousand Years of Unsustainable Growth: China's Environment from Archaic Times to the Present [Basis of the Annual Lecture of the Centre for Modern Chinese Studies at St Antony's College Oxford, 11 May 1994.]. *East Asian History* 6: 7–46.

Elvin, Mark. 2008. *The Retreat of the Elephants: An Environmental History of China.* New Haven: Yale University Press.

Endfield, Georgina H. 2012. The Resilience and Adaptive Capacity of Social-Environmental Systems in Colonial Mexico. *Proceedings of the National Academy of Sciences* 109(10): 3676–3681.

Endfield, Georgina H. 2014. Exploring Particularity: Vulnerability, Resilience and Memory in Climate Change Discourses. *Environmental History* 19(2): 303–310.

Endfield, Georgina H. and Veale, Lucy (eds.). 2018. *Cultural Histories, Memories and Extreme Weather: A Historical Geography Perspective.* New York: Routledge.

Engels, Friedrich. 1892. *The Condition of the Working Class in England in 1844 with Preface Written in 1892.* Translated by Wischnewetzsky, Florence Kelley. London: Swan Sonnenschein & Co.

Engle, Nathan L. 2011. Adaptive Capacity and Its Assessment. *Global Environmental Change* 21(2): 647–656.

Engler, Steven, Mauelshagen, Franz, Werner, Jürg & Luterbacher, Johannes. 2013. The Irish Famine of 1740–1741: Famine Vulnerability and "Climate Migration." *Climate of the Past* 9(3): 1161–1179.

Ensminger, Jean. 1992. *Making a Market: The Institutional Transformation of an African Society.* Cambridge: Cambridge University Press.

Eshghi, Kourosh & Larson, Richard C. 2008. Disasters: Lessons from the Past 105 Years. *Disaster Prevention and Management: An International Journal* 17(1): 62–82.

Farmer, Paul. 2012. *Haiti after the Earthquake.* New York: PublicAffairs.

Federico, Giovanni. 2010. *Feeding the World: An Economic History of Agriculture, 1800–2000.* Princeton & Oxford: Princeton University Press.

Felbermayr, Gabriel & Gröschl, Jasmin. 2011. Naturally Negative: The Growth Effects of Natural Disasters. *Journal of Development Economics* 111: 92–106.

Flinn, Michael W. 1974. The Stabilisation of Mortality in Pre-industrial Western Europe. *Journal of European Economic History* 3(2): 285–317.

Fogel, Robert W. 2003. *The Escape from Hunger and Premature Death, 1700–2100: Europe, America and the Third World.* Cambridge: Cambridge University Press.

Folke, Carl. 2006. Resilience: The Emergence of a Perspective for Socio-ecological Systems Analyses. *Global Environmental Change* 16(3): 253–267.

Folke, Carl, Carpenter, Stephen R., Walker, Brian *et al.* 2010. Resilience Thinking: Integrating Resilience, Adaptability and Transformability. *Ecology and Society* 15(4): Article 20.

Foreman-Peck, James. 2011. The Western-European Marriage Pattern and Economic Development. *Explorations in Economic History* 48(2): 292–309.

Forlin, Paolo & Gerrard, Christopher M. 2017. The Archaeology of Earthquakes: The Application of Adaptive Cycles to Seismically-Affected Communities in Late Medieval Europe. *Quaternary International* 446: 95–108.

Forlin, Paolo, Gerrard, Christopher M. & Petley, David N. 2016. Exploring Representativeness and Reliability for Late Medieval Earthquakes in Europe. *Natural Hazards* 84(3): 1625–1636.

Foster, Harold D. 2005. Assessing Disaster Magnitude: A Social Science Approach. *Professional Geographer* 28(3): 241–247.

Foster, John Bellamy. 1999. Marx's Theory of Metabolic Rift: Classical Foundations for Environmental Sociology. *American Journal of Sociology* 105 (2): 366–405.

Foucault, Michel. 1995. *Discipline and Punish: The Birth of the Prison.* New York: Vintage Books.

Frankema, Ewout & Masé, Aline. 2013. An Island Drifting Apart: Why Haiti Is Mired in Poverty While the Dominican Republic Forges Ahead. *Journal of International Development* 26(1): 128–148.

Franklin, Peter. 1986. Peasant Widows' "Liberation" and Remarriage before the Black Death. *Economic History Review* 39(2): 186–204.

Frerks, Georg & Bender, Stephen. 2004. Conclusion: Vulnerability Analysis as a Means of Strengthening Policy Formulation and Policy Practice, in Bankoff, Greg, Frerks, Georg & Hilhorst, Dorothea (eds.). *Mapping Vulnerability: Disasters, Development, and People.* London & Sterling: Routledge, 194–205.

Furió, Antonio. 2017. Inequality and Economic Development in Late Medieval Iberia, Catalonia and Valencia, 13th–16th century. Rural History Conference, Leuven, 11–14 September.

Fusco, Idamaria. 2017. The Importance of Prevention and Institutions: Governing the Emergency in the 1690–92 Plague Epidemic in the Kingdom of Naples. *Annales de Démographie Historique* 134(2): 95–123.

Füssel, Hans-Martin. 2007. Vulnerability: A Generally Applicable Conceptual Framework for Climate Change Research. *Global Environmental Change* 17(2): 155–167.

Galloway, James A. 2009. Storm Flooding, Coastal Defence and Land Use around the Thames Estuary and Tidal River c. 1250–1450. *Journal of Medieval History* 35(2): 171–188.

Galloway, Patrick R. 1988. Basic Patterns in Annual Variations in Fertility, Nuptiality, Mortality, and Prices in Pre-industrial Europe. *Population Studies: A Journal of Demography* 42(2): 275–303.

Geens, Sam. 2018. The Great Famine in the County of Flanders (1315–17): The Complex Interaction between Weather, Warfare, and Property Rights. *Economic History Review* 71(4): 1048–1072.

Gerber, Jean-David. 2017. New Commons and Resilience: Do CSR Schemes Compensate for "Commons Grabbing"? XVI Biennial IASC-Conference, Utrecht, 11 July.

Gerrard, Christopher J. & Petley, David N. 2013. A Risk Society? Environmental Hazards, Risk and Resilience in the Later Middle Ages in Europe. *Natural Hazards* 69(1): 1051–1079.

Gibbon, Edward. 1776. *The History of the Decline and Fall of the Roman Empire.* London: Strahan & Cadell.

Giddens, Anthony. 2013. *The Consequences of Modernity.* New York: John Wiley & Sons.

Gill, Tom, Steger, Brigitte & Slater, David H. 2013. *Japan Copes with Calamity: Ethnographies of the Earthquake, Tsunami, and Nuclear Disasters of March 2011.* Oxford: Peter Lang.

Glantz, Michael H. 1988. *Societal Responses to Regional Climatic Change: Forecasting by Analogy.* Boulder: Westview Press.

Glantz, Michael H. 1990. Does History Have a Future? Forecasting Climate Change Effects on Fisheries by Analogy. *Fisheries* 15(16): 39–44.

Glantz, Michael H. 2010. The Use of Analogies: In Forecasting Ecological and Societal Responses to Global Warming. *Environment: Science and Policy for Sustainable Development* 33(5): 10–33.

Gómez, J. M. & Verdú, M. 2017. Network Theory May Explain the Vulnerability of Medieval Human Settlements to the Black Death Pandemic. *Scientific Reports* 7: Article 43467.

Goubert, Pierre. 1970. Historical Demography and the Reinterpretation of Early Modern French History: A Research Review. *The Journal of Interdisciplinary History* 1(1): 37–48.

Goubert, Pierre. 1960. *Beauvais et le Beauvaisis de 1600 à 1730: Contribution à l'histoire sociale de la France au XVIIe siècle.* Paris: École pratique des Hautes-Études.

Green, Monica H. 2014. Taking "Pandemic" Seriously: Making the Black Death Global, in Green, Monica H. (ed.). *Pandemic Disease in the Medieval World: Rethinking the Black Death. The Medieval Globe,* Volume 1. Kalamazoo & Bradford: Arc Medieval Press, 27–61.

Green, Monica H. (ed.) 2014. *Pandemic Disease in the Medieval World: Rethinking the Black Death. The Medieval Globe,* Volume 1. Kalamazoo & Bradford: Arc Medieval Press.

Green, Monica. 2018. Black as Death. *Inference: International Review of Science* 4 (1), https://inference-review.com/article/black-as-death.

Groh, Dieter, Kempe, Michael & Mauelshagen, Franz. 2003. *Naturkatastrophen: Beiträge zu ihrer Deutung, Wahrnehmung und Darstellung in Text und Bild von der Antike bis ins 20. Jahrhundert.* Tübingen: Gunter Narr.

Grossman, Gene M. & Krueger, Alan B. 1995. Economic Growth and the Environment. *Quarterly Journal of Economics* 110(2): 353–377.

Guarino, Gabriel. 2006. Spanish Celebrations in Seventeenth-Century Naples. *The Sixteenth Century Journal* 37(1): 25–41.

Guha, Ramachandra. 2000. *The Unquiet Woods: Ecological Change and Peasant Resistance in the Himalaya.* Berkeley: University of California Press.

Guha, Ramachandra & Gadgil, Madhav. 1989. State Forestry and Social Conflict in British India. *Past & Present* 123(1): 141–177.

Guinnane, Timothy W. & Ogilvie, Sheilagh. 2014. A Two-Tiered Demographic System: "Insiders" and "Outsiders" in Three Swabian Communities, 1558-1914. *The History of the Family* 19(1): 77–119.

Gunderson, Lance H. 2000. Ecological Resilience: In Theory and Application. *Annual Review of Ecology and Systematics* 31(1): 425–439.

Gunderson, Lance H. & Holling, C. S. 2002. *Panarchy: Understanding Transformations in Human and Natural Systems.* Washington D.C.: Island Press.

Gutmann, Myron P. 1980. *War and Rural Life in the Early Modern Low Countries*. Assen: Van Gorcum.

Haemers, Jelle & Ryckbosch, Wouter. 2010. A Targeted Public: Public Services in Fifteenth-Century Ghent and Bruges. *Urban History* 37(2): 203–225.

Hahn, Micah B., Riederer, Anne M. & Foster, Stanley O. 2009. The Livelihood Vulnerability Index: A Pragmatic Approach to Assessing Risks from Climate Variability and Change – A Case Study in Mozambique. *Global Environmental Change* 19(1): 74–88.

Haldon, John, Elton, Hugh, Huebner, Sabine R. *et al.* 2018. Plagues, Climate Change, and the End of an Empire: A Response to Kyle Harper's *The Fate of Rome*. *History Compass* 16(12): Article e12508.

Haller, Tobias & Chabwela, Harry N. 2009. Managing Common Pool Resources in the Kafue Flats, Zambia: From Common Property to Open Access and Privatisation. *Development Southern Africa* 26(4): 555–567

Haller, Tobias. 2007. *Understanding Institutions and Their Links to Resource Management from the Perspective of New Institutionalism*, second edition. NCCR North–South Dialogue, no. 2: Bern: NCCR North–South.

Hamblyn, Richard. 2008. Notes from the Underground: Lisbon after the Earthquake. *Romanticism* 14(2): 108–118.

Hamlin, Christopher. 1998. *Public Health and Social Justice in the Age of Chadwick: Britain 1800–1854*. Cambridge: Cambridge University Press.

Hannaford, Matthew. 2018. Long-Term Drivers of Vulnerability and Resilience to Drought in the Zambezi–Save Area of Southern Africa, 1505–1830. *Global and Planetary Change* 166: 94–106.

Hannaford, Matthew. 2018. Pre-colonial South-East Africa: Sources and Prospects for Research in Economic and Social History. *Journal of Southern African Studies* 44(5): 771–792.

Hannaford, Matthew J. & Nash, David J. 2016. Climate, History, Society over the Last Millennium in Southeast Africa. *Wiley Interdisciplinary Reviews: Climate Change* 7(3): 370–392.

Hannaford, Matthew J., Jones, Julie M. & Bigg, Grant R. 2015. Early-Nineteenth-Century Southern African Precipitation Reconstructions from Ships' Logbooks. *The Holocene* 25(2): 379–390.

Hardin, Garrett. 1968. Tragedy of the Commons. *Science* 162(3859): 1243–1248.

Harper, Kyle. 2017. *The Fate of Rome: Climate, Disease, and the End of an Empire*. Princeton: Princeton University Press.

Hart, P. 't. 1993. Symbols, Rituals and Power: The Lost Dimensions of Crisis Management. *Journal of Contingencies and Crisis Management* 1(1): 36–50.

Hartman, Chester & Squires, Gregory (eds.). 2006. *There Is No Such Thing as a Natural Disaster: Race, Class, and Hurricane Katrina*. New York: Routledge.

Hatcher, John. 1977. *Plague, Population and the English Economy 1348–1530*. London: Macmillan & Co.

Hatcher, John. 1986. Mortality in the Fifteenth Century: Some New Evidence. *Economic History Review* 39(1): 19–38.

Hatcher, John. 2011. Unreal Wages: Long-Run Living Standards and the "Golden Age" of the Fifteenth Century, in Dodds, Ben & Liddy, Christian

D. (eds.). *Commercial Activity, Markets and Entrepreneurs in the Middle Ages: Essays in Honour of Richard Britnell*. Woodbridge: Boydell & Brewer, 1–24.

Hatcher, John & Bailey, Mark. 2001. *Modelling the Middle Ages: The History and Theory of England's Economic Development*. Oxford: Oxford University Press.

Healey, Jonathan. 2015. Famine and the Female Mortality Advantage: Sex, Gender and Mortality in Northwest England, c. 1590–1630. *Continuity and Change* 30(2): 153–192.

Heidinga, Hendrik A. 2010. The Birth of a Desert: The Kootwijkerzand, in Fanta, Josef & Siepel, Henk (eds.). *Inland Drift Sand Landscapes*. Zeist: KNNV Publishing, 65–80.

Herlihy, David. 1997. *The Black Death and the Transformation of the West*. Cambridge: Harvard University Press.

Herlihy, David & Klapisch-Zuber, Christiane. 1978. *Les Toscans et leurs familles: Une étude du catasto florentin de 1427*. Paris: Fondation nationale des Sciences Politiques & Éditions de l'École des Hautes Études en Sciences Sociales.

Hewitt, Kenneth. 1997. *Regions of Risk: A Geographical Introduction to Disasters*. Harlow: Longman.

Hill, Christopher V. 1997. *River of Sorrow: Environment and Social Control in Riparian North India, 1770–1994*. Ann Arbor: Association for Asian Studies.

Hillier, Debbie & Castillo, Gina E. 2013. *No Accident: Resilience and the Inequality of Risk*. Oxfam Briefing Paper 172. Cowley: Oxfam International.

Hindle, Steve. 2004. *On the Parish? The Micro-politics of Poor Relief in Rural England, 1550–1750*. Oxford: Oxford University Press.

Hinkel, Jochen. 2011. "Indicators of Vulnerability and Adaptive Capacity": Towards a Clarification of the Science–Policy Interface. *Global Environmental Change* 21(1): 198–208.

Hoddinott, John. 2006. Shocks and Their Consequences across and within Households in Rural Zimbabwe. *Journal of Development Studies* 42(2): 301–321.

Hodell, David A., Curtis, Jason H. & Brenner, Mark. 1995. Possible Role of Climate in the Collapse of Classic Maya Civilization. *Nature* 375(6530): 391–394.

Holling, Crawford S. 1973. Resilience and Stability of Ecological Systems. *Annual Review of Ecology and Systematics* 4(1): 1–23.

Holling, Crawford S. 2001. Understanding the Complexity of Economic, Ecological, and Social Systems. *Ecosystems* 4(5): 390–405.

Holmgren, Karin & Öberg, Helena. 2006. Climate Change in Southern and Eastern Africa During the Past Millennium and Its Implications for Societal Development. *Environment, Development and Sustainability* 8(1): 185–195.

Hooli, Lauri Johannes. 2016. Resilience of the Poorest: Coping Strategies and Indigenous Knowledge of Living with the Floods in Northern Namibia. *Regional Environmental Change* 16(3): 695–707.

Hornbeck, Richard. 2012. The Enduring Impact of the American Dust Bowl: Short- and Long-Run Adjustments to Environmental Catastrophe. *American Economic Review* 102(4): 1477–1507.

Hsiang, Solomon M., Burke, Marshall & Miguel, Edward. 2013. Quantifying the Influence of Climate on Human Conflict. *Science* 341(6151): Article 1235367.

Huang, Cheng, Li, Zhu, Wang, Meng & Martorell, Reynaldo. 2010. Early Life Exposure to the 1959–1961 Chinese Famine Has Long-Term Health Consequences. *The Journal of Nutrition* 140(10): 1874–1878.

Huffman, Thomas N. 2008. Climate Change during the Iron Age in the Shashe-Limpopo Basin, Southern Africa. *Journal of Archaeological Science* 35(7): 2032–2047.

Huhtamaa, Heli & Helama, Samuli. 2017. Reconstructing Crop Yield Variability in Finland: Long-Term Perspective of the Cultivation History on the Agricultural Periphery since AD 760. *The Holocene* 27(1): 3–11.

Humphries, Jane & Weisdorf, Jacob. 2015. The Wages of Women in England, 1260–1850. *Journal of Economic History* 75(2): 405–447.

Hunt, Margaret R. 2010. *Women in Eighteenth-Century Europe*. Harlow: Pearson Longman.

Ingram, M. J., Underhill, D. J. & Wigley, T. M. L. 1978. Historical Climatology. *Nature* 276(5686): 329–334.

IPCC. 2014. Summary for Policymakers, in Field, Christopher B., Barros, Vincente R., Dokken, David J. et al. (eds.). *Climate Change 2014: Impacts, Adaptation, and Vulnerability. Part A: Global and Sectoral Aspects. Contribution of Working Group II to the Fifth Assessment Report of the Intergovernmental Panel on Climate Change*. Cambridge: Cambridge University Press, 1–32.

Jackson Jr., Kennell A. 1976. The Family Entity and Famine among the Nineteenth-Century Akamba of Kenya: Social Responses to Environmental Stress. *Journal of Family History* 1(2): 193–216.

Jaeger, Carlo C., Renn, Ortwin, Rosa, Eugene A. & Webler, Thomas. 2001. *Risk, Uncertainty, and Rational Action*. London: Earthscan.

Jabukowski-Tiessen, Manfred. 2003. "Erschreckliche und unerhörte Wasserflut": Wahrnehmung and Deutung der Flutkatastrophe von 1634, in Jabukowski-Tiessen, Manfred & Lehmann, Hartmut (eds.). *Um Himmels Willen: Religion in Katastrophenzeiten*. Göttingen: Vandenhoeck & Ruprecht, 179–200.

Jakubowski-Tiessen, Manfred & Lehmann, Hartmut (eds.). 2003. *Um Himmels Willen: Religion in Katastrophenzeiten*. Göttingen: Vandenhoeck & Ruprecht.

Jodha, Narpat S. 2001. *Life on the Edge: Sustaining Agriculture and Community Resources in Fragile Environments*. Oxford: Oxford University Press.

Kahn, Matthew E. 2005. The Death Toll from Natural Disasters: The Role of Income, Geography, and Institutions. *Review of Economics and Statistics* 87(2): 271–284.

Kaika, Maria. 2017. *"Don't Call Me Resilient Again!"*: The New Urban Agenda as Immunology … or … What Happens When Communities Refuse To Be Vaccinated with "Smart Cities" and Indicators. *Environment and Urbanization* 29(1): 89–102.

Kaufmann, Maria, Lewandowski, Jakub, Choryński, Adam & Wiering, Mark. 2016. Shock Events and Flood Risk Management: A Media Analysis of the Institutional Long-Term Effects of Flood Events in the Netherlands and Poland. *Ecology and Society* 21(4): 51–66.

Kellenberg, Derek K. & Mobarak, Ahmed M. 2008. Does Rising Income Increase or Decrease Damage Risk from Natural Disasters? *Journal of Urban Economics* 63(3): 788–802.

Keller, A. Z., Meniconi, M., Al-Shammari, I. & Cassidy, K. 1997. Analysis of Fatality, Injury, Evacuation and Cost Data Using the Bradford Disaster Scale. *Disaster Prevention and Management* 6(1): 33–42.

Keller, A. Z., Wilson, H. C. & Al-Madhari, A. 1992. Proposed Disaster Scale and Associated Model for Calculating Return Periods for Disasters of Given Magnitude. *Disaster Prevention and Management: An International Journal* 1(1): 26–33.

Kelman, Ilan, Gaillard, J. C., Lewis, James & Mercer, Jessica. 2016. Learning from the History of Disaster Vulnerability and Resilience Research and Practice for Climate Change. *Natural Hazards* 82(1): 129–143.

Kelton, Paul. 2007. *Epidemics and Enslavement: Biological Catastrophe in the Native Southeast, 1492–1715*. Lincoln & London: University of Nebraska Press.

Kent, Randolph C. 1983. Reflecting upon a Decade of Disasters: The Evolving Response of the International Community. *International Affairs* 59(4): 693–711.

Keyzer, Maïka de. 2013. The Impact of Different Distributions of Power on Access Rights to the Common Wastelands: The Campine, Brecklands and Geest Compared. *Journal of Institutional Economics* 9(4): 517–542.

Keyzer, Maïka de. 2016. "All We Are Is Dust in the Wind": The Social Causes of a "Subculture of Coping" in the Late Medieval Coversand Belt. *Journal for the History of Environment and Society* 1(1): 1–35.

Keyzer, Maïka de & Bateman, Mark D. 2018. Late Holocene Landscape Instability in the Breckland (England) Drift Sands. *Geomorphology* 323: 123–134.

Keyzer, Maïka de. 2019.The Impact of Inequality on Social Vulnerability in Pre-modern Breckland. *Journal for the History of Environment and Society* 4.

King, Steven. 2011. Welfare Regimes and Welfare Regions in Britain and Europe, c. 1750s to 1860s. *Journal of Modern European History* 9(1): 44–67.

Kingston, Jeff. 2014. Mismanaging Risk and the Fukushima Nuclear Crisis, in Bacon, Paul & Hobson, Christopher (eds.). *Human Security and Japan's Triple Disaster: Responding to the 2011 Earthquake, Tsunami and Fukushima Nuclear Crisis*. Abingdon & New York: Routledge, Chapter 3.

Klein, Ira. 1984. When the Rains Failed: Famine, Relief, and Mortality in British India. *The Indian Economic and Social History Review* 21(2): 185–214.

Klein, Jørgen, Nash, David J., Pribyl, Kathleen, Endfield, Georgina H. & Hannaford, Matthew J. 2018. Climate, Conflict and Society: Changing Responses to Weather Extremes in Nineteenth Century Zululand. *Environment and History* 24(3): 377–401.

Klein, Naomi. 2007. *The Shock Doctrine: The Rise of Disaster Capitalism*. New York: Henry Holt and Company.

Klein Goldewijk, Kees. 2014. Environmental Quality since 1820, in Zanden, Jan Luiten van, Baten, Joerg, Mira d'Ercole, Marco *et al.* (eds.). *How Was Life?: Global Well-being since 1820*. Paris: OECD Publishing, 180–198.

Knowles, S. G. 2014. Learning from Disaster?: The History of Technology and the Future of Disaster Research. *Technology and Culture* 55(4): 773–784.

Koopmans, Joop W. 2014. The 1755 Lisbon Earthquake and Tsunami in Dutch News Sources: The Functioning of Early Modern News Dissemination, in Davies, Simon F. & Fletcher, Puck (eds.). *News in Early Modern Europe: Currents and Connections*. Leiden & Boston: Brill, 19–40.

Kosminsky, E. A. 1956. *Studies in the Agrarian History of England in the Thirteenth Century*. Oxford: Basil Blackwell.

Kowaleski, Maryanne. 2013. Gendering Demographic Change in the Middle Ages, in Bennett, Judith & Karras, Ruth (eds.). *The Oxford Handbook of Women and Gender in Medieval Europe*. Oxford: Oxford University Press, 181–196.

Krüger, Fred, Bankoff, Greg, Cannon, Terry, Orlowski, Benedikt & Schipper, E. Lisa F. (eds.). *Cultures and Disasters: Understanding Cultural Framings in Disaster Risk Reduction*. London & New York: Routledge.

Kutalek, Ruth, Wang, Shiyong, Fallah, Mosoka, Wesseh, Chea Sanford & Gilbert, Jeffrey. 2015. Ebola Interventions: Listen to Communities. *The Lancet Global Health* 3(3): Article E131.

Laborda Pemán, Miguel & Moor, Tine de. 2013. A Tale of Two Commons: Some Preliminary Hypotheses on the Long-Term Development of the Commons in Western and Eastern Europe, 11th–19th Centuries. *The International Journal of the Commons* 7(1): 7–33.

Lamb, Hubert H. 1972/1977. *Climate: Past, Present and Future*, Volumes 1 & 2. Milton Park & New York: Routledge.

Lamb, Hubert H. 1995. *Climate, History and the Modern World*. London & New York: Methuen.

Lambrecht, Thijs. 2003. Reciprocal Exchange, Credit and Cash: Agricultural Labour Markets and Local Economies in the Southern Low Countries during the Eighteenth Century. *Continuity and Change* 18(2): 237–261.

Lambrecht, Thijs. 2017. The Harvest of the Poor? Comparative Perspectives on Gleaning and Poor Relief in France and England, c. 1540–c. 1840. Rural History Conference, Leuven, 11–14 September.

Landers, John. 1986. Mortality, Weather and Prices in London 1675–1825: A Study of Short-Term Fluctuations. *Journal of Historical Geography* 12(4): 347–364.

Le Roy Ladurie, Emmanuel. 1967. *Histoire du climat depuis l'an mil*. Paris: Flammarion.

Le Roy Ladurie, Emmanuel. 1971. *Times of Feast, Times of Famine: A History of Climate since the Year 1000*. Translated by Bray, Barbara. Garden City: Doubleday.

Le Roy Ladurie, Emmanuel & Baulant, Micheline. 1980. Grape Harvests from the Fifteenth through the Nineteenth Centuries. *Journal of Interdisciplinary History* 10(4): 839–849.

Lee, Jeffrey A. & Gill, Thomas E. 2015. Multiple Causes of Wind Erosion in the Dust Bowl. *Aeolian Research* 19(Part A): 15–36.

Leeuwen, Marco H. D. van. 2016. *Mutual Insurance 1550–2015: From Guild Welfare and Friendly Societies to Contemporary Micro-insurers*. London: Palgrave Macmillan.

Levi, Giovanni. 2003. Aequitas vs Fairness: Reciprocità ed equità fra età moderna ed età contemporanea. *Rivista di Storia Economica* 19(2): 195–204.

Lewis, Carenza. 2016. Disaster Recovery: New Archaeological Evidence for the Long-Term Impact of the "Calamitous" fourteenth century. *Antiquity* 351(90): 777–797.

Lewis, Mary. 2016. Work and the Adolescent in Medieval England AD 900–1550: The Osteological Evidence. *Medieval Archaeology* 60(1): 138–171.

Li, Lillian M. 1991. Life and Death in a Chinese Famine: Infanticide as a Demographic Consequence of the 1935 Yellow River Flood. *Comparative Studies in Society and History* 33(3): 466–510.

Lindstrom, David P. & Berhanu, Betemariam. 1999. The Impact of War, Famine, and Economic Decline on Marital Fertility in Ethiopia. *Demography* 36(2): 247–261.

Lis, Catharina & Soly, Hugo. 1979. *Poverty and Capitalism in Pre-industrial Europe*. Hassocks: Harvester Press.

Little, Lester K. 2011. Plague Historians in Lab Coats. *Past & Present* 213(1): 267–290.

Livi Bacci, Massimo. 1978. *La société italienne devant les crises de mortalité*. Florence: Dipartimento di Statistica, Università di Firenze (Conférences au Collège de France).

Logan, John R. & Molotch, Harvey L. 1987. *Urban Fortunes: The Political Economy of Place*. Berkeley: University of California Press.

Long, Pamela O. 2018. *Engineering the Eternal City: Infrastructure, Topography, and the Culture of Knowledge in Late Sixteenth-Century Rome*. Chicago: The University of Chicago Press.

Lübken, Uwe & Mauch, Christof. 2011. Uncertain Environments: Natural Hazards, Risk and Insurance in Historical Perspective. *Environment and History* 17(1): 1–12.

Luterbacher, Jürg, Dietrich, Daniel, Xoplaki, Elena, Grosjean, Martin & Wanner, Heinz. 2004. European Seasonal and Annual Temperature Variability, Trends, and Extremes since 1500. *Science* 303(5663): 1499–1503.

Luzzadder-Beach, Sheryl, Beach, Timothy P. & Dunning, Nicholas P. 2012. Wetland Fields as Mirrors of Drought and the Maya Abandonment. *Proceedings of the National Academy of Sciences* 109(10): 3646–3651.

Lynteris, Christos. 2018. Suspicious Corpses: Body Dumping and Plague in Colonial Hong Kong, in Lynteris, Christos & Evans, N. (eds.). *Histories of Post-Mortem Contagion: Medicine and Biomedical Sciences in Modern History*. Cham: Palgrave Macmillan, 109–133.

MacKinnon, Danny & Driscoll Derickson, Kate. 2012. From Resilience to Resourcefulness: A Critique of Resilience Policy and Activism. *Progress in Human Geography* 37(2): 253–270.

Madhav, Nita, Oppenheim, Ben, Gallivan, Mark et al. 2017. Pandemics: Risks, Impacts and Mitigation, in Jamison, D. T., Gelbrand, H., Horton, S. et al. (eds.). *Disease Control Priorities: Improving Health and Reducing Poverty*. Washington: The International Bank for Reconstruction and Development/ The World Bank, Chapter 17.

Ma, James. 2000. Path Dependence in Historical Sociology. *Theory and Society* 29 (4): 507–548.
Mahoney, James. 2002. *The Legacies of Liberalism: Path Dependence and Political Regimes in Central America*. Baltimore: John Hopkins University Press.
Mahoney, James. 2010. *Colonialism and Postcolonial Development: Spanish America in Comparative Perspective*. New York: Cambridge University Press.
Malanima, Paolo. 2008. The Economic Consequences of the Black Death. Paper given at the International Conference Roma-Anacapri, Rome, 9–11 October.
Malanima, Paolo. 2009. *Pre-modern European Economy: One Thousand Years (10th–19th Centuries)*. Leiden & Boston: Brill.
Manley, Gordon. 1972. *Climate and the British Scene*. London: Collins.
Maramai, Alessandra, Brizuela, Beatriz & Graziani, Laura. 2014. The Euro-Mediterranean Tsunami Catalogue. *Annals of Geophysics* 57(4): Article S0435.
Marfany, Julie. 2017. Quantifying the Unquantifiable? Informal Charity in Southern Europe, Seventeenth and Eighteenth Centuries. Rural History Conference, Leuven, 11–14 September.
Marincioni, F. 2001. A Cross-cultural Analysis of Natural Disaster Response: The Northwest Italy Floods of 1994 Compared to the US Midwest Floods of 1993. *International Journal of Mass Emergencies and Disasters* 19(2): 209–236.
Martinez-Alier, Joan. 2002. *The Environmentalism of the Poor: A Study of Ecological Conflicts and Valuation*. Cheltenham: Edward Elgar.
Mauelshagen, Franz. 2007. Flood Disasters and Political Culture at the German North Sea Coast: A Long-Term Historical Perspective. *Historical Social Research* 32(3): 133–144.
Mauelshagen, Franz. 2011. Sharing the Risk of Hail: Insurance, Reinsurance and the Variability of Hailstorms in Switzerland, 1880–1932. *Environment and History* 17(1): 171–191.
Mauelshagen, Franz. 2014. Redefining Historical Climatology in the Anthropocene. *The Anthropocene Review* 1(2): 171–204.
Mauelshagen, Franz, 2015. Defining Catastrophes, in Gerstenberger, Katharina & Nusser, Tanja (eds.). *Catastrophe and Catharsis: Perspectives on Disasters and Redemption in German Culture and Beyond*. Rochester: Boydell & Brewer, 172–190.
Mauelshagen, Franz & Pfister, Christian. 2010. Vom Klima zur Gesellschaft: Klimageschichte im 21. Jahrhundert, in Welzer, Harald, Soeffner, Hans-Georg & Giesecke, Dana (eds.). *Klimakulturen: Soziale Wirklichkeiten im Klimawandel*. Frankfurt am Main: Campus, 241–269.
McCann, James. 2001. Maize and Grace: History, Corn and Africa's New Landscapes, 1500–1999. *Comparative Studies in Society and History* 43(2): 246–272.
McCloskey, D. N. 1976. English Open Fields as Behavior towards Risk. *Research in Economic History* 1(2): 124–171.
McCloskey, Donald N. 1991. The Prudent Peasant: New Findings on Open Fields. *Journal of Economic History* 51(2): 343–355.
McNeill, John R. 2010. *Mosquito Empires: Ecology and War in the Greater Caribbean, 1620–1914*. Cambridge: Cambridge University Press.

McNeill, John R. 2010. Sustainable Survival, in McAnany, Patricia A. & Yoffee, Norman. *Questioning Collapse: Human Resilience, Ecological Vulnerability and the Aftermath of Empire.* Cambridge: Cambridge University Press, 355–366.

McNeill, William H. 1992. *The Global Condition: Conquerors, Catastrophes and Community.* Princeton: Princeton University Press.

Meuvret, Jean. 1969. Les oscillations des prix des céréales aux XVIIe et XVIIIe siècles en Angleterre et dans les pays du Bassin parisien. *Revue d'histoire moderne et contemporaine* 16(4): 540–554.

Middleton, Guy D. 2012. Nothing Lasts Forever: Environmental Discourses on the Collapse of Past Societies. *Journal of Archaeological Research* 20(3): 257–307.

Mikhail, Alan. 2011. *Nature and Empire in Ottoman Egypt: An Environmental History.* Cambridge: Cambridge University Press.

Milanovic, Branko, 2016. *Global Inequality: A New Approach for the Age of Globalization.* Cambridge: Belknap Press.

Milanovic, Branko. 2016. Income Inequality Is Cyclical. *Nature* 537(7621): 479–482.

Miller, Joseph C. 1982. The Significance of Drought, Disease and Famine in the Agriculturally Marginal Zones of West-Central Africa. *The Journal of African History* 23(1): 17–61.

Mitchell, Timothy. 2002. *Rule of Experts: Egypt, Techno-politics, Modernity.* Berkeley: University of California Press.

Moor, Tine de. 2008. The Silent Revolution: A New Perspective on the Emergence of Commons, Guilds, and Other Forms of Corporate Collective Action in Western Europe. *International Review of Social History* 52(16): 179–212.

Moor, Tine de. 2009. Avoiding Tragedies: A Flemish Common and Its Commoners under the Pressure of Social and Economic Change during the Eighteenth Century. *Economic History Review* 62(1): 1–22.

Moor, Tine de. 2010. Participating Is More Important Than Winning: The Impact of Socio-economic Change on Commoners' Participation in Eighteenth- and Nineteenth-Century Flanders. *Continuity and Change* 25(3): 405–433.

Moor, Tine de. 2014. Single, Safe, and Sorry? Explaining the Early Modern Beguine Movement in the Low Countries. *Journal of the Family* 39 (1): 3–21.

Moor, Tine de. 2015. *The Dilemma of the Commoners: Understanding the Use of Common-Pool Resources in Long-Term Perspective.* Cambridge: Cambridge University Press.

Moor, Tine de & Zanden, Jan Luiten van. 2010. Girl Power: The European Marriage Pattern and Labour Markets in the North Sea Region in the Late Medieval and Early Modern Period. *Economic History Review* 63(1): 1–33.

Moore, Jason W. 2000. Environmental Crises and the Metabolic Rift in World-Historical Perspective. *Organization & Environment* 12(3): 123–157.

Moore, Jason W. (ed.). 2016. *Anthropocene or Capitalocene?: Nature, History, and the Crisis of Capitalism.* Oakland: PM Press.

Moore, Jason W. 2017. The Capitalocene, Part I: On the Nature and Origins of Our Ecological Crisis. *Journal of Peasant Studies* 44(3): 594–630.

Morera, Raphael. 2010. Environmental Change and Globalization in Seventeenth-Century France: Dutch Traders and the Draining of French Wetlands (Arles, Petit Poitou). *International Review of Social History* 55(18): 79–101.

Mukerji, Chandra. 2007. Stewardship Politics and the Control of Wild Weather: Levees, Seawalls, and State Building in 17th-Century France. *Social Studies of Science* 37(1): 127–133.

Mulcahy, Matthew. 2008. *Hurricanes and Society in the British Greater Caribbean, 1624–1783*. Baltimore: Johns Hopkins University Press.

Munck, Bert de. 2009. Fiscalizing Solidarity (from Below): Poor Relief in Antwerp Guilds: Between Community Building and the Public Service, in Heijden, Manon van der (ed.). *Serving the Urban Community: The Rise of Public Facilities in the Low Countries*. Amsterdam: Akant, 168–193.

Munck, Bert de. 2018. *Guilds, Labour and the Urban Body Politic: Fabricating Community in the Southern Netherlands, 1300–1800*. New York: Routledge.

Murphy, Neil. 2013. Plague Ordinances and the Management of Infectious Diseases in Northern French Towns, c. 1450–c. 1560, in Clark, Linda & Rawcliffe, Carole (eds.). *The Fifteenth Century XII: Society in an Age of Plague*. Woodbridge: Boydell & Brewer, 139–160.

Naphy, William G. & Spicer, Andrew. 2001. *The Black Death: A History of Plagues, 1345–1730*. Stroud: Tempus.

Naphy, William G. 2002. *Plagues, Poisons, and Potions: Plague-Spreading Conspiracies in the Western Alps, c. 1530–1640*. Manchester: Manchester University Press.

Nash, David J. & Adamson, George C. D. 2014. Recent Advances in the Historical Climatology of the Tropics and Subtropics. *Bulletin of the American Meteorological Society* 95(1): 131–146.

Nash, David J., Pribyl, Kathleen, Klein, Jørgen et al. 2014. Tropical Cyclone Activity over Madagascar during the Late Nineteenth Century. *International Journal of Climatology* 35(11): 3249–3261.

Nash, David J., Pribyl, Kathleen, Klein, Jørgen et al. 2016. Seasonal Rainfall Variability in Southeast Africa during the Nineteenth Century Reconstructed from Documentary Sources. *Climatic Change* 134(4): 605–619.

Nee, Victor & Ingram, Paul. 1998. Embeddedness and Beyond: Institutions, Exchange, and Social Structure, in Brinton, Mary C. & Nee, Victor (eds.). *The New Institutionalism in Sociology*. Stanford: Stanford University Press.

Nelkin, Dorothy & Gilman, Sander L. 1988. Placing Blame for Devastating Disease. *Social Research* 55(3): 361–378.

Nelson, Donald R., Adger, W. Neil & Brown, Katrina. 2007. Adaptation to Environmental Change: Contributions of a Resilience Framework. *Annual Review of Environment and Resources* 32: 395–419.

Nelson, Margaret C., Ingram, Scott E., Dugmore, Andrew J. et al. 2016. Climate Challenges, Vulnerabilities, and Food Security. *Proceedings of the National Academy of Sciences* 113(2): 298–303.

Newson, Linda A. 1993. The Demographic Collapse of Native Peoples of the Americas, 1492–1650, in Bray, Warwick (ed.). *The Meeting of Two Worlds: Europe and the Americas, 1492–1650*. Oxford, Oxford University Press, 247–288.

Noble, Ian R., Huq, Saleemul, Anokhin, Yuri A. et al. 2014. Adaptation Needs and Options, in Field, Christopher B., Barros, Vincente R., Dokken, David J. et al. (eds.). *Climate Change 2014: Impacts, Adaptation, and Vulnerability. Part A: Global and Sectoral Aspects. Contribution of Working Group II to the Fifth Assessment Report of the Intergovernmental Panel on Climate Change*. Cambridge: Cambridge University Press, 833–868.

North, Douglass C. 1990. *Institutions, Institutional Change and Economic Performance*. Cambridge: Cambridge University Press.

Ó Gráda, Cormac. 2007. Ireland's Great Famine: An Overview, in Ó Gráda, Cormac, Paping, Richard, & Vanhaute, Eric (eds.). *When the Potato Failed: Causes and Effects of the "Last" European Subsistence Crisis, 1845–1850*. Turnhout: Brepols, 43–57.

Ó Gráda, Cormac. 2010. *Famine: A Short History*. Princeton: Princeton University Press.

Ó Gráda, Cormac & Chevet, Jean-Michel, 2002. Famine and Market in Ancien Régime France. *Journal of Economic History* 62(3): 706–733.

Ó Gráda, Cormac & Mokyr, Joel. 2002. What Do People Die of during Famines: The Great Irish Famine in Comparative Perspective. *European Review of Economic History* 6(3): 339–363.

Ó Gráda, Cormac, Paping, Richard, & Vanhaute, Eric (eds.). 2007. *When the Potato Failed: Causes and Effects of the "Last" European Subsistence Crisis, 1845–1850*. Turnhout: Brepols.

O'Connor, Daniel, Boyle, Philip, Ilcan, Suzan & Oliver, Marcia. 2017. Living with Insecurity: Food Security, Resilience, and the World Food Programme (WFP). *Global Science Policy* 17(1): 3–20.

O'Keefe, Phil, Westgate, Ken & Wisner, Ben. 1976. Taking the Naturalness out of Natural Disasters. *Nature* 260(5552): 566–567.

Ogilvie, Sheilagh. 2007. "Whatever Is, Is Right"? Economic Institutions in Preindustrial Europe. *Economic History Review* 60(4): 649–684.

Ogilvie, Sheilagh. 2019. *The European Guilds: An Economic Analysis*. Princeton: Princeton University Press.

Ogilvie, Sheilagh. & Carus, A. W. 2014. Institutions and Economic Growth in Historical Perspective, in Aghion, Philippe & Durlauf, Steven N. (eds.). *Handbook of Economic Growth*, Volume 2. Amsterdam: Elsevier, 403–513.

Oliver-Smith, Anthony. 1994. Peru's Five-Hundred-Year Earthquake: Vulnerability in Historical Context, in Varley, Anne (ed.). *Disasters, Development, and Environment*. Chichester: Wiley, 3–48.

Oliver-Smith, Anthony. 1996. Anthropological Research on Hazards and Disasters. *Annual Review of Anthropology* 25: 303–328.

Onacker, Eline van. 2019. Social Vulnerability and Social Structures in Sixteenth-Century Flanders: A Micro-level Analysis of Household Grain Shortage during the Crisis of 1556/57. *Continuity and Change* 34(1): 91–115.

Onacker, Eline van & Masure, Hadewych. 2015. Unity in Diversity: Rural Poor Relief in the Sixteenth-Century Southern Low Countries. *Tijdschrift voor Sociale en Economische Geschiedenis* 12(4): 59–88.

Oosten, Roos van. 2016. The Dutch Great Stink: The End of the Cesspit Era in the Pre-industrial Towns of Leiden and Haarlem. *European Journal of Archaeology* 19(4): 704–727.

Oosthuizen, Susan. 2017. Water Management for Sustainable Pastoral Productivity on Early Medieval Commons: The Peat Wetlands of Eastern England, *c.* 600–900 AD. Rural History Conference, Leuven, 11–14 September.

Oosthuizen, Susan & Willmoth, Frances. 2009. *Drowned and Drained: Exploring Fenland Records and Landscape*. Cambridge: Institute of Continuing Education, University of Cambridge.

Ostrom, Elinor. 2015. *Governing the Commons: The Evolution of Institutions for Collective Action*. Cambridge: Cambridge University Press.

Overlaet, Kim. 2014. Replacing the Family? Beguinages in Early Modern Western European Cities: An Analysis of the Family Networks of Beguines Living in Mechelen (1532–1591). *Continuity and Change* 29(3): 325–347.

Overton, Mark. 1989. Weather and Agricultural Change in England, 1660–1739. *Agricultural History* 63(2): 77–88.

PAGES2k Consortium. 2017. A Global Multiproxy Database for Temperature Reconstructions of the Common Era. *Scientific Data* 4: Article 170088.

Pamuk, Şevket. 2007. The Black Death and the Origins of the "Great Divergence" across Europe, 1300–1600. *European Review of Economic History* 11(3): 289–317.

Parker, Geoffrey. 2008. Crisis and Catastrophe: The Global Crisis of the Seventeenth Century Reconsidered. *The American Historical Review* 113(4): 1053–1079.

Parker, Geoffrey. 2014. *The Global Crisis: War, Climate, and Catastrophe in the Seventeenth Century World*. New Haven: Yale University Press.

Pastore, Alessandro. 1991. *Crimine e giustizia in tempo di peste nell'Europa moderna*. Rome & Bari: Laterza.

Pellecchia, Umberto, Crestani, Rosa, Decroo, Tom, Berg, Rafael van den & Al-Kourdi, Yasmine. 2015. Social Consequences of Ebola Containment Measures in Liberia. *PLoS ONE* 10(12): Article e0143036.

Pelling, Mark. 2003. *The Vulnerability of Cities: Natural Disasters and Social Resilience*. London & Sterling: Earthscan.

Pelling, Mark. 2011. *Adaptation to Climate Change: From Resilience to Transformation*. London: Routledge.

Perdue, Peter C. 1982. Official Goals and Local Interests: Water Control in the Dongting Lake Region during the Ming and Qing Periods. *Journal of Asian Studies* 41(4): 747–765.

Pereira, Alvaro S. 2009. The Opportunity of a Disaster: The Economic Impact of the 1755 Lisbon Earthquake. *Journal of Economic History* 69(2): 466–499.

Perry, Roland W. & Quarantelli, Enrico Louis. 2005. *What Is a Disaster?: New Answers to Old Questions*. Philadelphia: Xlibris.

Pfister, Christian. 2007. Climatic Extremes, Recurrent Crises and Witch Hunts: Strategies of European Societies in Coping with Exogenous Shocks in the Late Sixteenth and Early Seventeenth Centuries. *The Medieval History Journal* 10 (1–2): 1–41.

Pfister, Christian. 2009. Die "Katastrophenlücke" des 20. Jahrhunderts und der Verlust traditionalen Risikobewusstseins. *GAIA* 18(3): 239–246.

Pfister, Christian. 2010. The Vulnerability of Past Societies to Climatic Variation: A New Focus for Historical Climatology in the 21st Century. *Climatic Change* 100(1): 25–31.

Pfister, Christian & Brázdil, Rudolf. 2006. Social Vulnerability to Climate in the Little Ice Age: An Example from Central Europe in the Early 1770s. *Climate of the Past* 2(2): 115–129.

Pielke Jr., Roger, Prins, Gwyn, Rayner, Steve & Sarewitz, Daniel. 2007. Climate Change 2007: Lifting the Taboo on Adaptation. *Nature* 445 (7128): 597–598.

Pierik, Harm Jan, Lanen, Rowin J. van, Gouw-Bouman, Marjolein T. I. J. *et al.* 2018. Controls on Late-Holocene Drift-Sand Dynamics: The Dominant Role of Human Pressure in the Netherlands. *The Holocene* 28(9): 1361–1381.

Pierson, Paul. 2000. Increasing Returns, Path Dependence, and the Study of Politics. *The American Political Science Review* 94(2): 251–267.

Piervitali, E. & Colacino, M. 2001. Evidence of Drought in Western Sicily during the Period 1565–1915 from Liturgical Offices. *Climatic Change* 49(1–2): 225–238.

Piketty, Thomas. 2014. *Capital in the Twenty-First Century*. Cambridge: Harvard University Press.

Pirngruber, Reinhard. 2012. Plagues and Prices: Locusts, in Baker, Heather D. & Jursa, Michael (eds.). *Documentary Sources in Ancient Near Eastern and Greco-Roman History: Methodology and Practice*. Oxford: Oxbow Books, 163–186.

Platt, Harold L. 2005. *Shock Cities: The Environmental Transformation and Reform of Manchester and Chicago*. Chicago: University of Chicago Press.

Polanyi, Karl. 1944. *The Great Transformation*. New York: Farrar & Rinehart, Inc.

Prak, Maarten R., Lis, Catharina, Lucassen, Jan & Soly, Hugo. 2006. *Craft Guilds in the Early Modern Low Countries: Work, Power and Representation*. Aldershot: Ashgate.

Pretty, Jules N. 1990. Sustainable Agriculture in the Middle Ages: The English Manor. *Agricultural History Review* 38(1): 1–19.

Pribyl, Kathleen. 2017. *Farming, Famine and Plague: The Impact of Climate in Late Medieval England*. Cham: Springer International Publishing.

Prieto, María del Rosario & Herrera, Ricardo García. 2009. Documentary Sources from South America: Potential for Climate Reconstruction. *Palaeogeography, Palaeoclimatology, Palaeoecology* 281(3–4): 196–209.

Pritchard, Sara B. 2012. An Envirotechnical Disaster: Nature, Technology, and Politics at Fukushima. *Environmental History* 17(2): 219–243.

Putnam, Robert. 2000. *Bowling Alone: The Collapse and Revival of American Community*. New York: Simon & Schuster.

Quarantelli, Enrico Louis. 1987. Disaster Studies: An Analysis of the Social Historical Factors Affecting the Development of Research in the Area. *International Journal of Mass Emergencies and Disasters* 5(3): 285–310.

Quarantelli, Enrico Louis. 1998. Disaster Planning, Emergency Management, and Civil Protection: The Historical Development and Current Characteristics of Organized Efforts to Prevent and to Respond to Disasters. *DRC Preliminary Papers* 227.

Quarantelli, Enrico Louis (ed.). 1998. *What Is a Disaster? Perspectives on the Question*. London & New York: Routledge.

Radkau, Joachim. 2008. *Nature and Power: A Global History of the Environment*. Cambridge: Cambridge University Press.

Rawcliffe, Carole. 2006. *Leprosy in Medieval England*. Woodbridge: Boydell Press.

Rawcliffe, Carole. 2013. *Urban Bodies: Communal Health in Late Medieval English Towns and Cities*. Woodbridge: Boydell Press.

Raworth, Kate. 2017. *Doughnut Economics: Seven Ways to Think Like a 21st-Century Economist*. New York: Random House.

Razi, Zvi. 1993. The Myth of the Immutable English Family. *Past & Present* 140 (1): 3–44.

Redman, Charles L. & Kinzig, Ann P. 2003. Resilience of Past Landscapes: Resilience Theory, Society, and the *Longue Durée*. *Ecology and Society* 7(1): Article 14.

Reed Jr., Adolph L. 2005. The Real Divide. *The Progressive* 69(11): 27–32.

Reid, Julian. 2012. The Disastrous and Politically Debased Subject of Resilience. *Development Dialogue* 58(1): 67–79.

Reis, Jaime. 2017. Deviant Behaviour? Inequality in Portugal 1565–1770. *Cliometrica* 11(3): 297–319.

Renaud, Fabrice G., Dun, Olivia, Warner, Koko & Bogardi, Janos. 2011. A Decision Framework for Environmentally Induced Migration. *International Migration* 49(1): e5–e29.

Rennie, J. J., Lawrimore, J. H., Gleason, B. E., *et al.* 2014. The International Surface Temperature Initiative Global Land Surface Databank: Monthly Temperature Data Release Description and Methods. *Geoscience Data Journal* 1(2): 75–102.

Rheinheimer, Martin. 2003. Mythos Sturmflut: Der Kampf gegen das Meer und die Suche nach Identität. *Demokratische Geschichte* 15: 9–58.

Ribot, Jesse. 2014. Cause and Response: Vulnerability and Climate in the Anthropocene. *Journal of Peasant Studies* 41(5): 667–705.

Riede, Felix. 2014. Towards a Science of Past Disasters. *Natural Hazards* 71(1): 335–362.

Riede, Felix. 2017. Past-Forwarding Ancient Calamities. Pathways for Making Archeology Relevant in Disaster Risk Reduction Research. *Humanities* 6(4): Article 79.

Riede, Felix, Bazely, Oliver, Newton, Anthony J. & Lane, Christine S. 2011. A Laacher See-Eruption Supplement to Tephrabase: Investigating Distal Tephra Fallout Dynamics. *Quaternary International* 246(1–2): 134–144.

Rigby, Stephen H. 2000. Gendering the Black Death: Women in Later Medieval England. *Gender and History* 12(3): 745–754.

Rivers, J. P. W. 1982. Women and Children Last: An Essay on Sex Discrimination in Disasters. *Disasters* 6(4): 256–267.

Robson, Elly. 2019. *Improvement and Environmental Conflict in the Northern Fens, 1560–1665*. PhD Thesis: Cambridge University.

Rockström, Johan, Steffen, Will, Noone, Kevin *et al*. 2009. Planetary Boundaries: Exploring the Safe Operating Space for Humanity. *Ecology and Society* 14(2): Article 32.

Rodgers, James E. T. 1884. *Six Centuries of Work and Wages: The History of English Labour*. London: Swan Sonnenschein.

Rodrigo, Fernando S. & Barriendos, M. 2008. Reconstruction of Seasonal and Annual Rainfall Variability in the Iberian Peninsula (16th–20th Centuries) from Documentary Data. *Global and Planetary Change* 63(2–3): 243–257.

Rodríguez, Havidán, Quarantelli, Enrico L. & Dynes, Russel R. (eds.). 2007. *Handbook of Disaster Research*. New York: Springer.

Rohland, Eleonora. 2012. Earthquake versus Fire: The Struggle over Insurance in the Aftermath of the 1906 San Francisco Disaster, in Janku, Andrea, Schenk, Gerrit & Mauelshagen, Franz (eds.) *Historical Disasters in Context: Science, Religion and Politics*. New York: Routledge, 174–194.

Rohland, Eleonora. 2018. Adapting to Hurricanes: A Historical Perspective on New Orleans from Its Foundation to Hurricane Katrina, 1718–2005. *Wiley Interdisciplinary Reviews: Climate Change* 9(1): Article e488.

Rohland, Eleonora. 2019. *Changes in the Air: Hurricanes in New Orleans*. New York: Berghahn Books.

Rohr, Christian. 2003. Man and Natural Disaster in the Late Middle Ages: The Earthquake in Carinthia and Northern Italy on 25 January 1348 and its Perception. *Environment and History* 9(2): 127–149.

Ronsmans, Carine, Chowdhury, Mahbub E., Dasgupta, Sushil. K. *et al*. 2010. Effect of Parent's Death on Child Survival in Rural Bangladesh: A Cohort Study. *The Lancet* 375(9730): 2024–2031.

Roosbroeck, Filip van. 2015. Experts, experimenten en veepestbestrijding in de Oostenrijkse Nederlanden, 1769–1785. *Tijdschrift voor Geschiedenis* 128(1): 23–43.

Roosbroeck, Filip van & Sundberg, Adam. 2017. Culling the Herds? Regional Divergences in Rinderpest Mortality in Flanders and South Holland, 1769–1785. *Tijdschrift voor Sociale en Economische Geschiedenis* 14(3): 31–55.

Roosen, Joris & Curtis, Daniel R. 2018. Dangers of Noncritical Use of Historical Plague Data. *Emerging Infectious Diseases* 24(1): 103–110.

Roosen, Joris & Curtis, Daniel R. 2019. The "Light Touch" of the Black Death in the Southern Netherlands: An Urban Trick? *Economic History Review* 72(1): 32–56.

Rose, Colin. 2018. Plague and Violence in Early Modern Italy. *Renaissance Quarterly* 71(3): 1000–1035.

Rosenthal, J. L. & Wong, R. B. 2011. *Before and beyond Divergence: The Politics of Economic Change in China and Europe*. Cambridge: Harvard University Press.

Ross, Eric B. 1995. Syphilis, Misogyny, and Witchcraft in 16th-Century Europe. *Current Anthropology* 36(2): 333–337.

Rotberg, Robert I. & Rabb, Theodore K. 1981. *Climate and History: Studies in Interdisciplinary History*. Princeton: Princeton University Press.

Rubin, Olivier. 2009. The Merits of Democracy in Famine Protection – Fact or Fallacy? *European Journal of Development Research* 21(5): 699–717.

Ryser, Sarah. 2017. Moroccan Regeneration: Government Land-Deal as a Catalytic Converter for Neo-Liberal and State-Driven Social Development. XVI Biennial IASC-Conference, Utrecht, 12 July.

Sachs, Jeffrey & Malaney, Pia. 2002. The Economic and Social Burden of Malaria. *Nature* 415(6872): 680–685.

Sartori, Giovanni. 1991. Comparing and Miscomparing. *Journal of Theoretical Politics* 3(3): 243–257.

Schama, Simon. 1988. *The Embarrassment of Riches: An Interpretation of Dutch Culture in the Golden Age: An Interpretation of Dutch Culture in the Golden Age*. Berkeley: University of California Press.

Scheffer, Marten. 2009. *Critical Transitions in Nature and Society*. Princeton: Princeton University Press.

Scheffer, Marten, Bavel, Bas van, Leemput, Ingrid A. van de & Nes, Egbert H. van. 2017. Inequality in Nature and Society. *Proceedings of the National Academy of Sciences* 114(50): 13154–13157.

Scheidel, Walter. 2017. *The Great Leveler: Violence and the Global History of Inequality from the Stone Age to the Present*. Princeton & Oxford: Princeton University Press.

Schenk, Gerrit Jasper (ed.). 2017. *Historical Disaster Experiences: Towards a Comparative and Transcultural History of Disasters across Asia and Europe*. New York: Springer.

Schipper, E. L. F., Merli, Claudia & Nunn, Patrick D. 2014. How Religion and Beliefs Influence Perceptions of and Attitudes towards Risk, in Cannon, Terry, Schipper, Lisa, Bankoff, Greg & Krüger, Fred (eds.). *World Disasters Report 2014: Focus on Cultures and Risk*. International Federation of Red Cross and Red Crescent Societies, 37–63. Available at www.ifrc.org/Global/Documents/Secretariat/201410/WDR%202014.pdf, consulted 7 November 2019.

Schneider, François, Kallis, Giorgos & Martinez-Alier, Joan. 2010. Crisis or Opportunity? Economic Degrowth for Social Equity and Ecological Sustainability. Introduction to this Special Issue. *Journal of Cleaner Production* 18(6): 511–518.

Schofield, Phillipp R. 2007. The Social Economy of the Medieval Village in the Early Fourteenth Century. *Economic History Review* 61(1): 38–63.

Schubert, Siegfried D., Suarez, Max J., Pegion, Philip J., Koster, Randal D., Bacmeister, Julio T. 2004. On the Cause of the 1930s Dust Bowl. *Science* 5665 (303): 1855–1859.

Sella, Domenico. 1991. Coping with Famine: The Changing Demography of an Italian Village in the 1590s. *The Sixteenth Century Journal* 22(2): 185–197.

Sellars, Emily A. & Alix-Garcia, Jennifer. 2018. Labor Scarcity, Land Tenure, and Historical Legacy: Evidence from Mexico. *Journal of Development Economics* 135: 504–516.

Sen, Amartya. 1998. Mortality as an Indicator of Economic Success and Failure. *Economic Journal* 108(446): 1–25.

Sen, Amartya. 2001. *Development as Freedom*. Oxford: Oxford University Press.
Sen, Amartya K. 1981. *Poverty and Famines: An Essay on Entitlement and Deprivation*. Oxford: Clarendon Press.
Serrão, José V. & Santos, Rui. 2013. Land Policies and Land Markets: Portugal, Late Eighteenth and Early Nineteenth Century, in Béaur, Gérard, Schofield, Philipp R., Chevet, Jean-Michel, & Picazo, María T. P. (eds.). *Property Rights, Land Markets and Economic Growth in the European Countryside (Thirteenth–Twentieth Centuries)*. Turnhout: Brepols, 317–341.
Shapin, Steven. 2007. Science and the Modern World, in Hackett, Edward, Amsterdamska, Olga, Lynch, Michael E. & Wajcman, Judy (eds.). *The Handbook of Science and Technology Studies*. Cambridge: MIT Press, 433–448.
Sharma, Sanjay. 2001. *Famine, Philanthropy and the Colonial State: North India in the Early Nineteenth Century*. Oxford: Oxford University Press.
Shaw, Ian. 2003. *The Oxford History of Ancient Egypt*. Oxford: Oxford University Press.
Shefer-Mossensohn, Miri. 2017. Rethinking Historiography and Sources: A Historiography of Epidemics in the Islamic Mediterranean, in Varlik, Nükhet (ed.). *Plague and Contagion in the Islamic Mediterranean: New Histories of Disease in Ottoman Society*. Kalamazoo & Bradford: Arc Humanities Press, 3–26.
Shepherd, Andrew, Mitchell, Tom, Lewis, Kirsty et al. 2013. *The Geography of Poverty, Disasters and Climate Extremes in 2030*. London: Overseas Development Institute.
Shirk, Melanie V. 1984. Violence and the Plague in Aragon, 1348–1351. *Journal of the Rocky Mountain Medieval and Renaissance Association* 5: 31–39.
Sidky, Homayun. 2010. *Witchcraft, Lycanthropy, Drugs and Disease: An Anthropological Study of the European Witch-Hunts*. Eugene: Wipf and Stock Publishers.
Singh, Praveen. 2008. The Colonial State, Zamindars and the Politics of Flood Control in North Bihar (1850–1945). *Indian Economic and Social History Review* 45(2): 239–259.
Singleton, John. 2015. Using the Disaster Cycle in Economic and Social History, in Brown, A. T., Burn, Andy & Doherty, Rob (eds.). *Crises in Economic and Social History: A Comparative Perspective*. Woodbridge: Boydell & Brewer.
Slavin, Philip. 2013. Market Failure during the Great Famine in England and Wales (1315–1317). *Past & Present* 222(1): 9–49.
Slavin, Philip. 2016. Epizootic Landscapes: Sheep Scab and Regional Environment in England in 1279–1280. *Landscapes* 17(2): 156–170.
Slavin, Philip. 2019. *Experiencing Famine in Fourteenth-Century Britain*. Turnhout: Brepols.
Smith, Keith & Petley, David N. 2009. *Environmental Hazards: Assessing Risk and Reducing Disaster*. Abingdon: Routledge.
Smith, S. D. 2011. Volcanic Hazard in a Slave Society: The 1812 Eruption of Mount Soufrière in St Vincent. *Journal of Historical Geography* 37(1): 55–67.
Snowden, Frank M. 1991. Cholera in Barletta 1910. *Past & Present* 132(1): 67–103.

Soens, Tim. 2005. Explaining Deficiencies of Water Management in the Late Medieval Flemish Coastal Plain (13th–16th Centuries). *Jaarboek voor Ecologische Geschiedenis*: 35–61.

Soens, Tim. 2006. Polders zonder poldermodel? Een onderzoek naar de rol van inspraak en overleg in de waterstaat van de laatmiddeleeuwse Vlaamse kustvlakte (1250–1600). *Tijdschrift voor Sociale en Economische Geschiedenis* 3 (4): 3–36.

Soens, Tim. 2009. *De spade in de dijk?: Waterbeheer en rurale samenleving in de Vlaamse kustvlakte (1280–1580)*. Ghent: Academia Press.

Soens, Tim. 2011. Floods and Money: Funding Drainage and Flood Control in Coastal Flanders from the Thirteenth to the Sixteenth Centuries. *Continuity and Change* 26(3): 333–365.

Soens, Tim. 2013. Flood Security in the Medieval and Early Modern North Sea Area: A Question of Entitlement? *Environment and History* 19(2): 209–232.

Soens, Tim. 2018. Resilient Societies, Vulnerable People: Coping with North Sea Floods Before 1800. *Past & Present* 241(1): 143–177.

Soens, Tim, Tys, Dries & Thoen, Erik. 2014. Landscape Transformation and Social Change in the North Sea Polders, the Example of Flanders (1000–1800 AD). *Siedlungforschung: Archäologie, Geschichte, Geographie* 31: 133–160.

Sommer, Matthew H. 2015. *Polyandry and Wife-Selling in Qing Dynasty China: Survival Strategies and Judicial Interventions*. Oakland: University of California Press.

Speakman, John. 2013. Sex- and Age-Related Mortality Profiles during Famine: Testing the "Body Fat" Hypothesis. *Journal of Biosocial Science* 45(6): 823–840.

Squatriti, Paolo. 2010. The Floods of 589 and Climate Change at the Beginning of the Middle Ages: An Italian Microhistory. *Speculum* 85(4): 799–826.

Steffen, Will, Crutzen, Paul. J & McNeill, John R. 2007. The Anthropocene: Are Humans Now Overwhelming the Great Forces of Nature. *AMBIO: A Journal of the Human Environment* 36(8): 614–621.

Steinberg, Ted. 2006. *Acts of God: The Unnatural History of Natural Disaster in America*. New York: Oxford University Press.

Sudmeier-Rieux, Karen I. 2014. Resilience: An Emerging Paradigm of Danger or of Hope? *Disaster Prevention and Management* 23(1): 67–80.

Swyngedouw, Erik. 2006. Circulations and Metabolisms: (Hybrid) Natures and (Cyborg) Cities. *Science as Culture* 15(2): 105–121.

Tainter, Joseph A. 1988. *The Collapse of Complex Societies*. Cambridge: Cambridge University Press.

Tarr, Joel A. 1996. *The Search for the Ultimate Sink: Urban Pollution in Historical Perspective*. Akron: University of Akron Press.

Theilmann, John & Cate, Frances. 2007. A Plague of Plagues: The Problem of Plague Diagnosis in Medieval England. *Journal of Interdisciplinary History* 37 (3): 371–393.

Thoen, Erik & Soens, Tim. 2008. The Social and Economic Impact of Central Government Taxation on the Flemish Countryside (end 13th–18th centuries), in Cavaciocchi, Simonetta (ed.). *La fiscalità nell'economia europea secc. XIII–XVII*. Florence: Firenze University Press, 957–971.

Tierney, Kathleen J. 2006. Social Inequality, Hazards, and Disasters, in Daniels, Ronald J., Kettl, Donald F. & Kunreuther, Howard (eds.). *On Risk and Disaster: Lessons from Hurricane Katrina*. Philadelphia: University of Pennsylvania Press, 109–128.

Tierney, Kathleen J. 2007. From the Margins to the Mainstream? Disaster Research at the Crossroads. *Annual Review of Sociology* 33(1): 503–525.

Titow, J. 1960. Evidence of Weather in the Account Rolls of the Bishopric of Winchester 1209-1350. *Economic History Review* 12(3): 360–407.

Titow, Jan. 1970. Le climat à travers les rôles de comptabilité de l'évêché de Winchester (1350–1450). *Annales: Histoire, Sciences Sociales* 25(2): 312–350.

Tol, Richard S. J. & Wagner, Sebastian. 2010. Climate Change and Violent Conflict in Europe over the Last Millennium. *Climatic Change* 99(1–2): 65–79.

United Nations Economic Commission for Africa. 1995. *A Symposium: Famine in Ethiopia: Learning from the Past to Prepare for the Future*. Addis Ababa: UN Economic Commission for Africa.

United Nations Office for Disaster Risk Reduction. 2015. *UNISDR Annual Report 2015*. Geneva: UNISDR, www.unisdr.org/files/48588_unisdrannualreport2015evs.pdf.

Vanhaute, Eric & Lambrecht, Thijs. 2011. Famine, Exchange Networks and the Village Community: A Comparative Analysis of the Subsistence Crises of the 1740s and the 1840s in Flanders. *Continuity and Change* 26(2): 155–186.

Vardi, Liana. 1993. Construing the Harvest: Gleaners, Farmers, and Officials in Early Modern France. *The American Historical Review* 98(5): 1424–1447.

Vaughan, Megan. 1985. Famine Analysis and Family Relations: 1949 in Nyasaland. *Past & Present* 108(1): 177–205.

Vernon, James. 2005. The Ethics of Hunger and the Assembly of Society: The Techno-politics of the School Meal in Modern Britain. *American Historical Review*, 110(3): 693–725.

Voigtländer, Nico & Voth, Hans-Joachim. 2013. How the West "Invented" Fertility Restriction. *American Economic Review* 103(6): 2227–2264.

Voigtländer, Nico & Voth, Hans-Joachim. 2013. The Three Horsemen of Riches: Plague, War, and Urbanization in Early Modern Europe. *Review of Economic Studies* 80(2): 774–811.

Vreyer, Philippe de & Nilsson, Björn. 2019. When Solidarity Fails: Heterogeneous Effects on Children from Adult Deaths in Senegalese Households. *World Development* 114: 73–94.

Vries, Jan de. 1980. Measuring the Impact of Climate on History: The Search for Appropriate Methodologies. *Journal of Interdisciplinary History* 10(4): 599–630.

Waal, Alex de. 1989. Famine Mortality: A Case Study of Darfur, Sudan 1984–5. *Population Studies* 43(1): 5–24.

Waal, Alex de. 2009. *Famine Crimes: Politics and the Disaster Relief Industry in Africa*. Bloomington: Indiana University Press.

Wahl, T., Haigh, I. D., Nicholls, R. J. et al. 2017. Understanding Extreme Sea Levels for Broad-Scale Coastal Impact and Adaptation Analysis. *Nature Communications* 8: Article 16075.

Wallerstein, Immanuel. 2011. *The Modern World-System I: Capitalist Agriculture and the Origins of the European World-Economy in the Sixteenth Century*. Berkeley: University of California Press.
Walter, John & Schofield, Roger. 1989. Famine, Disease and Crisis Mortality in Early Modern Society, in Walter, John & Schofield, Roger (eds.). *Famine, Disease and the Social Order in Early Modern Society*. Cambridge: Cambridge University Press, 1–74.
Warde, Paul. 2015. Global Crisis or Global Coincidence? *Past & Present* 228(1): 287–301.
Watts, Michael J. & Bohle, Hans G. 1993. The Space of Vulnerability: The Causal Structure of Hunger and Famine. *Progress in Human Geography* 17(1): 43–67.
Weick, Karl E. 1993. The Collapse of Sensemaking in Organizations: The Mann Gulch Disaster. *Administrative Science Quarterly* 38(4): 628–652.
Weintritt, Otfried. 2009. The Floods of Baghdad: Cultural and Technological Responses, in Mauch, Christof & Pfister, Christian (eds.). *Natural Disasters, Cultural Responses: Case Studies towards a Global Environmental History*. Plymouth: Lexington Books, 165–182.
Wesselink, Anna J. 2007. Flood Safety in the Netherlands: The Dutch Response to Hurricane Katrina. *Technology in Society* 29(2): 239–247.
Wetter, Oliver & Pfister, Christian. 2011. Spring–Summer Temperatures Reconstructed for Northern Switzerland and Southwestern Germany from Winter Rye Harvest Dates, 1454–1970. *Climate of the Past* 7(4): 1307–1326.
Wheatcroft, S. G. 2020. Societal Responses to Food Shortages and Famine in Russia and China, in Dijkman, Jessica & Leeuwen, Bas van (eds.). *An Economic History of Famine Resilience*. London: Routledge, 182–202.
White, Gilbert Fowler. 1945. *Human Adjustment to Floods: A Geographical Approach to the Flood Problem in the United States*. PhD Thesis: The University of Chicago.
White, Sam. 2012. *The Climate of Rebellion in the Early Modern Ottoman Empire*. New York: Cambridge University Press.
Wickham, Chris. 2010. *The Inheritance of Rome: A History of Europe from 400 to 1000*. London: Penguin Books.
Wigley, T. M. L., Ingram, M. J. & Farmer, G. (eds.). 1981. *Climate and History: Studies in Past Climates and Their Impact on Man*. Cambridge: Cambridge University Press.
Wilkinson, Annie & Fairhead, James. 2016. Comparison of Social Resistance to Ebola Response in Sierra Leone and Guinea Suggests Explanations Lie in Political Configurations not Culture. *Critical Public Health* 27(1): 14–27.
Will, Pierre-Etienne & Wong, Bin. 1991. *Nourish the People: The State Civilian Granary System in China*. Ann Arbor: University of Michigan Press.
Wilson Bowers, Kristy. 2013. *Plague and Public Health in Early Modern Seville*. Rochester: University of Rochester Press.
Wise, Russ, Fazey, Ioan, Stafford Smith, Mark *et al.* 2014. Reconceptualising Adaptation to Climate Change as Part of Pathways of Change and Response. *Global Environmental Change* 28(1): 325–336.

Wisner, Ben, Blaikie, Piers, Cannon, Terry & Davis, Ian. 2004. *At Risk: Natural Hazards, People's Vulnerability and Disasters*, second edition. London & New York: Routledge.
Wisner, Ben. 1993. Disaster Vulnerability: Scale, Power and Daily Life. *GeoJournal* 30(2): 127–140.
Wong, Koon-Kwai & Zhao, Xiaobin. 2001. Living with Floods: Victims' Perceptions in Beijiang, Guandong, China. *Area* 33(2): 190–201.
Wood, Harry O. & Neumann, Frank. 1931. Modified Mercalli Intensity Scale of 1931. *Seismological Society of America Bulletin* 21(4): 277–283.
Worster, Donald. 1979. *Dust Bowl: The Southern Plains in the 1930s*. Oxford: Oxford University Press.
Wray, S. Kelly. 2004. Boccaccio and the Doctors: Medicine and Compassion in the Face of Plague. *Journal of Medieval History* 30(3): 301–322.
Wray, S. Kelly. 2009. *Communities and Crisis: Bologna during the Black Death*. Leiden & Boston: Brill.
Wrigley, Edward A. & Schofield, Roger S. 1989. *The Population History of England 1541–1871*. Cambridge: Cambridge University Press.
Yang, Bin & Cao, Shuji. 2006. Cadres, Grain and Sexual Abuse in Wuwei County, Mao's China. *Journal of Women's History* 28(2): 33–57.
Yaussy, Samantha L., DeWitte, Sharon N. & Redfern, Rebecca C. 2016. Frailty and Famine: Patterns of Mortality and Physiological Stress among Victims of Famine in Medieval London. *American Journal of Physical Anthropology* 160(2): 272–283.
Yue, Ricci P. H., Lee, Harry F. & Wu, Connor Y. H. 2016. Navigable Rivers Facilitated the Spread and Recurrence of Plague in Pre-industrial Europe. *Scientific Reports* 6: Article 34867.
Zanden, Jan Luiten van. 1999. The Paradox of the Marks: The Exploitation of Commons in the Eastern Netherlands, 1250–1850. *The Agricultural History Review* 47(2): 125–144.
Zarulli, Virginia, Barthold Jones, Julia A., Oksuzyan, Anna et al. 2018. Women Live Longer Than Men Even during Severe Famines and Epidemics. *Proceedings of the National Academy of Sciences* 155(4): E832–E840.
Zhang, David D., Brecke, Peter, Lee, Harry F., He, Yuan-Qing & Zhang, Jane. 2007. Global Climate Change, War, and Population Decline in Recent Human History. *Proceedings of the National Academy of Sciences* 104(49): 19214–19219.
Zhang, David D., Lee, Harry F., Wang, Cong et al. 2011. The Causality Analysis of Climate Change and Large-Scale Human Crisis. *Proceedings of the National Academy of Sciences* 108(42): 17296–17301.
Zhang, David D., Zhang, Jane, Lee, Harry F. & He, Yuan-qing. 2007. Climate Change and War Frequency in Eastern China over the Last Millennium. *Human Ecology* 35(4): 403–414.
Zhao, Zhongwei & Reimondos, Anna. 2012. The Demography of China's 1958–61 Famine: A Closer Examination. *Population* 67(2): 281–308.
Zijdeman, Richard L. & Ribeiro da Silva, Filipa. 2014. Life Expectancy since 1820, in Zanden, Jan Luiten van, Baten, Joerg, Mira d'Ercole, Marco et al. (eds.). *How Was Life?: Global Well-being since 1820*. Paris: OECD Publishing, 101–116.

Index

adaptation
  connection to vulnerability, 67
  development of the concept, 38
  negative elements, 38
  positive elements, 38
  temporality of, 38, 39
adaptive capacity. *See* adaptation
adaptive cycle, 39
Adger, W.N., 35
aeolian soils, 75
Age of Disaster, 9, 185
agricultural technology, 78
Alfani, Guido, 137
American Dust Bowl. *See* Dust Bowl
American Great Plains, 75. *See* Dust Bowl
Anthropocene, 9, 159, 160, 161, 164, 168
  concept of, 160
  Great Acceleration, 160
Armenia
  Spitak earthquake of 1988, 134
Ash, Eric, 113

Babylonian Astronomical Diaries. *See* historical sources
bacterium *Vibrio cholerae*. *See* cholera outbreaks
Baghdad
  thirteenth-century water management, 89
Bangladesh
  famine of 1974, 125
Bankoff, Greg, 40, 75, 103, 183
Bavel, Bas van, 166
Beck, Ulrich, 40, 42, 167
Behringer, Wolfgang, 138
Beijiang basin flood
  reliance on family, 87
Bengal famine of 1943, 14
  role of the market, 88
Biraben, Jean-Noël, 46, 58
Black Death, 7, 8, 11, 13, 14, 16, 21, 26, 55, 64, 68, 69, 103, 179. *See also* epidemics

  and gender inequality, 156
  and inequality, 155
  broader assessment, 69
  compassion and cohesion, 140
  disparate effects, 147
  disparate population recovery, 150
  diversity of outcomes, 21
  Golden Age of Labour, 14
  long-term positive effects, 147
  migration to cities, 131
  mortality, 64
  persecution of Jews, 140
  redistributive anomaly, 156
  social tensions, 138
Blaikie, Piers, 11
Borsch, Stuart, 21
Brazil
  indigenous marriage rules, 149
Brecke, Peter, 58. *See also* Conflict Catalog
burial records, 48, 64, 65, 124. *See also* historical sources
  reconstructing mortality, 65

Campbell, Bruce, 17, 146
Canal du Midi, 90
capitalism
  and cyclicality, 166
  and hybridization, 165
  and societal change, 152
  changing human–nature interaction, 165
  frontier zone vulnerability, 166
  nonlinear development, 166
Capitalocene, 164, 165
Catania
  earthquake of 1693, 125, 132
cattle plague
  and intensification, 82
China
  famine of 1958–61, 124, 129
  floods and economic effects, 148
  Mandate of Heaven, 101, 177
  top-down interventions, 106

Chinese gazetteers. *See* historical sources
cholera outbreaks
  bottom-up resistance, 141
  in the nineteenth century, 79, 114
  sanitary reform, 114
chronicles. *See also* historical sources
classquakes. *See* earthquakes
climate change
  and disaster risk, 161, 164
Climatic Research Unit (CRU), 13
commons
  and inequality, 95
  and vulnerability, 94
  collective agriculture, 93
  disintegration, 95
  in Northwestern Europe, 92
  shared non-arable resources, 94
  water boards, 96
communist revolutions
  and food crises, 152
Conflict Catalog, 58
coordination systems, 87
  and resources, 85
  collective action. *See also* commons
  the family, 88
  the market, 89
  the state, 91
coping mechanisms, 5, 49, 91, 118, 161, 185, 186, 187
COVID-19 pandemic, 10, 108

D'Souza, Rohan, 180
Darfur
  famine of 1984–5, 125
Davis, Mike, 166, 180
de Keyzer, Maïka, 119
Di Tullio, Matteo, 119
disaster classification, 22
  consequences, 24
  natural and human-made, 22
  speed of unfolding, 23
  type of event, 23
disaster experts
  and biopolitics, 114
  co-evolvement with the state, 113
  modern era dominance, 114
  rise of, 113
disaster impacts, 5, 62
  capital destruction, 5, 132, 134
  economic crisis, 137
  erosion of societal stability, 5
  falling yields, 5, 136
  fixed values and thresholds, 63
  land loss, 134

mortality, 48, 63, 125. *See also* selective mortality
  scapegoating and blame, 140
  structural and behavioral disturbance, 63
disaster management cycle, 31
  and decision-making, 33
  development of, 32
  limitations, 33
  long-term perspective, 32
  mitigation and adaptation, 32
disaster preconditions, 74
  culture gaps, 104
  economic development, 82
  economic structures, 84
  hazard recurrence. *See also* regions of risk
  religious beliefs, 101
  risks of mitigation measures, 79
  technology and infrastructure, 77, 78
  war and conflict, 85
Disaster Research Center (DRC), 11
disaster responses, 105
  cooperation from below, 109, 141
  top-down and bottom-up, 106
disaster studies
  and anthropologists, 11
  and climate change, 11
  and disaster management, 185
  and eurocentrism, 184
  and forecasting, 169
  and history, 1, 169, 171, 182
  at the local level, 184
  Cold War period, 11
  field, 1, 14, 171
  focus on victims, 186, 187
  global direction, 69
  in the 1990s, 11
  interdisciplinary research, 54, 182
  limitations, 54
  potential for historians, 2, 173, 178, 179, 181
disasters
  and modernity, 10
  and origin myths, 178
  and poverty, 69, 82
  and social vulnerability. *See* vulnerability
  concept of, 30
  difference to hazards, 31
  naturalness of, 33
diseases
  colonialism and capitalism, 166
  in the Caribbean, 166
  in the Southeast of the United States, 166
  nineteenth-century reforms, 151
  references in historical sources, 64
diversity of outcomes, 2, 8, 158, 163

Index

demographic, 7
economic, 7
institutional responses, 3, 181
lasting economic divergence, 137
social actors, 4
Dust Bowl, 5, 6, 8, 12, 75, 83, 132
and migration, 132
land loss, 132
long-term economic depression, 137

earthquakes
and capital destruction, 134
contextualization of, 70
positive effects, 146
Ebert, John David, 160
Ebola outbreaks
bottom-up resistance, 110, 141
in Western Africa, 84, 104
Egypt
Nile Delta engineering, 115
water management, 98, 148
Egyptian Nilometers, 46. See also historical sources
Emergency Events Database (EM-DAT), 161
data coverage, 161
Endfield, Georgina, 36
Engels, Friedrich, 18
England
Breckland sand drifts, 99
population reconstruction, 65
sheep scab epizootic of 1279–80, 136
Enlightenment, 9, 28, 167
Ensminger, Jean, 118
Environmental Kuznets Curve (EKC), 175
concept, 176
limitations, 176
epidemics, 6
and inheritance, 155
and long-term effects, 147
and migration, 131
and public health, 151
and redistribution, 155
and religion in Western Europe, 101
and social control, 109
and social unrest, 138
and top-down interventions, 109
blame and scapegoating, 137
bottom-up defiance, 109, 110
bottom-up resistance, 141
burial practices, 109, 141
elite repression, 138, 153
indiscriminate mortality, 129
marriage spike, 130
mortality, 64

negotiated balance, 140
productivity and human capital, 147
reconstruction, 63
escape from disaster, 175
through economic growth, 177
Ethiopia
crisis of the 1980s, 129
European Marriage Pattern, 87
explanatory frameworks, 16
institutional approach, 21
Malthusian approach, 18
Marxist approach, 19
modernization approach, 20

famines
and democracy, 89
and economic diversification, 82
and Malthusian pressures, 69
and migration, 149
and poor relief organizations, 92
anticipation in Africa, 111
charity in India, 93
comparative approach, 69
diversity of outcomes, 158
Great Famine of 1315–17, 68
in Ethiopia and Sudan, 90, 151
sub-Saharan Africa, 8
informal relief, 93, 158
Malthusian pressures, 175
nuptiality and fertility, 149
postponing marriage, 129
related diseases, 26, 148
selective mortality, 126
floods
along the North Sea coast, 8, 76, 79, 81, 102, 125, 152
in England, 76
protection through mounds, 79
water management systems, 81
Friedman, Milton, 152, 153

Germany
failure to adapt, 103
flood of 1634, 101
Fukushima disaster implications, 160
Gibbon, Edward, 142
and moral decay, 142
Glantz, Michael H., 169
Global South, 34, 69, 81, 178, 183
Great Dhaka Flood Protection Project
top-down intervention side effects, 108
Great Famine of 1315–17, 68, 69
impact of the Flemish–French war, 85
in England, 100

Great Famine of Ireland, 69, 175
  contextualization, 70
  grain exports to England, 84
  potato dependency, 83
Great Transition, 146
Guatemala
  earthquake of 1976, 81
Guha, Ramachandra, 19
guilds. *See* commons

Haiti
  colonial path dependency, 172
  earthquake of 2010, 26, 134, 172
Hardin, Garrett, 94
hazards. *See* shocks
  concept of, 30
  coping with, 4
  nature of, 30
  proneness to, 76
Herlihy, David, 17
Hewitt, Kenneth, 75
histoire événementielle. *See also* historiography of disasters
historical catalogs. *See also* historical sources
historical data
  applying social science methods, 71
  colonial records, 58, 60
  further development, 60
  in the digital age, 57
  key questions, 60
  paleoclimate proxy data, 169
  source criticism, 58
  unevenness of, 58, 70, 161
historical sources
  and colonialism, 50
  contextualization of data, 57
  instrumental recording, 50
  integration of sources, 55
  lack of, 8, 135
  limitations of, 8, 124, 149, 158
  non-documentary evidence, 55
  types of, 53
  under-representation, 124, 183
historiography of disasters, 15, 179
  drivers of change, 179
  environmental determinism, 178
  focus on events, 69
  historical climatology, 13
  in the post-World War II period, 12
  in the pre-industrial period, 12
  socio-economic history, 14
history as a laboratory, 43
  focus on institutions, 174
  future goals, 183
  identifying patterns, 173

practical demands, 43
social contextualization, 174
testing hypotheses, 43, 183
HIV/AIDS, 68, 147
  blame and scapegoating, 137
Holling, Buzz, 35
Holocene, 75, 76
hurricane Katrina, 22, 82
  and education policy change, 153
  and family size, 87
  and poverty, 96, 180
  Bipartisan Report, 113
  political and social bias, 181
  selective mortality, 125

Ibn Khaldun, 142
  and moral decay, 142
India, 50
  Bhopal gas tragedy, 40, 167
  droughts and imperialism, 180
  floods and colonial capitalism, 180
  plague in Surat in 1994, 112, 138
Indian Ocean
  tsunami of 2004, 168
Industrial Revolution, 9, 10, 18, 82, 159, 164, 167
inequality
  and colonialism, 118, 180
  and disaster preconditions, 157
  and indirect vulnerability, 99
  and taxation, 158
  and war, 155
  benefits of more equity, 118
  beyond wealth distribution, 156
  constrained disaster response, 117
  disaster preconditions, 179
  enhanced hazard response, 117
  long-term trends, 155
  of income, 97
  reciprocal agreements, 99
informal solidarity, 49
institutions
  and change, 86
  and resilience, 120
  colonial path dependency, 121
  coordination systems. *See* coordination systems
  culture and religion, 122
  flexibility and adaptation, 118
  formation of, 86
  hazard-orientated, 85
  marginalizing indigenous beliefs, 121
  path dependency, 86, 120, 173
Integrated History and Future of People on Earth (IHOPE) network, 170, 173

Index

methodological approach, 171
Intergovernmental Panel on Climate
  Change (IPCC), 37, 161
  concept of risk, 162
Italy
  Black Death in, 7
  plague epidemic of 1629–30, 7, 26, 137
  plague epidemic of 1656–8, 137
  reciprocity, 119

Japan
  Fukushima disaster, 8, 134, 154, 160
  Great Kantō earthquake of 1923, 134
  Nuclear and Industry Safety Authority
    (NISA), 154

Kant, Immanuel
  and the Lisbon earthquake of 1755, 167
Kelman, Ilan, 163
Klein, Naomi, 152
Kosminsky, E. A., 18
Kuznets, Simon, 175

Lamb, Hubert, 13
Le Roy Ladurie, Emmanuel, 12, 13,
  139, 178
Leiden
  epidemics and inequality, 117
Lisbon
  earthquake of 1755, 8, 11, 23, 27,
    28, 49, 50, 90, 132, 134, 135, 145,
    153, 168
Little Ice Age, 38, 55, 76
  and societal collapse, 143
  and witch hunting, 139
  diversity of outcomes, 163
  regional diversity, 77
Livi Bacci, Massimo, 65
long-term disaster effects, 145
  demographic changes, 150
  economic divergence, 146
  gradual change, 153
  reconstruction and reform, 152
Low Countries
  accommodation paradigm, 116
  Campine area sand drifts, 102,
    108, 120
  Delta Works, 116, 176
  flood protection investments, 83
  floods of 1953, 116, 176
  Kootwijk dust bowl, 133, 148
  land loss and recovery, 134
  plague data, 58
  poor relief and local elites, 119
  technological lock-in, 116
  water management, 98, 99, 102

Malawi. *See* Southern Nyasaland
Manchuria. *See* Manchurian plague of 1911
Manchurian plague of 1911, 50
Manley, Gordon, 13
Marquis of Pombal, 28, 49, 153
  institutional reforms, 135
Marx, Karl, 19, 165
Mauelshagen, Franz, 102, 103
McCloskey, Donald, 41
McNeill, William H., 177
media coverage. *See also* historical sources
  source critique, 50
Medieval Warm Optimum, 76
memory and learning, 110
  de-learning triggers, 112
  preconditions, 112, 181
metabolic rift, 19
  concept of, 165
methodologies, 61
  analog, 170
  comparative approach, 68, 73, 174
  practical challenges to comparisons, 72
  reconstruction, 62
Mexico
  epidemic in the 1570s, 157
  pre-Hispanic droughts, 102
Milanovic, Branko, 7, 154, 158
Modified Mercalli Intensity scale, 24
Moore, Jason, 83, 164, 165, 166

New Orleans
  absence of memory and learning, 112
Northern Italy
  famine of the 1590s, 82
  flood protection, 117
  top-down famine intervention, 106

O'Keefe, Phil, 81
optically stimulated luminescence (OSL)
  dating, 56
Ostrom, Elinor, 21
Overton, Mark, 178

Peru
  colonial path dependency, 172
  earthquake of 1970, 171
peste. *See* plague
Pfister, Christian, 13, 138, 139
Philippines, 54
  bottom-up mitigation, 111
  coping with risks, 103
Piketty, Thomas, 157
plague
  in Naples, 109
  in Ragusa, 109, 151
  terminology of, 64. *See also* epidemics

Polanyi, Karl, 164, 166
poor relief organizations
    financial context, 93
population recovery, 132
    migration, 132
    nuptiality and fertility, 130, 149
    societal shifts, 131
Postan, Michael, 16
Pressure and Release (PAR) model, 66
Pritchard, Sara, 160
public sphere
    Enlightenment disaster interest, 168
Puerto Rico
    hurricane Maria in 2018, 135

Quarantelli, Enrico, 11

Radkau, Joachim, 168
Red Cross, 91
regions of risk, 75, 77, 101
    and adaptation, 102
    embeddedness of risk, 101
    failure to adapt, 103
    subcultures of coping, 103
resilience. *See* vulnerability
    conceptual critique, 37
    development of the concept, 36
rinderpest, 49, 90
    in the Southern Netherlands, 49, 107
Ring of Fire, 76
risk
    and fear, 167
    and 'othering', 41, 76, 186
    as a physical construct, 162
    aspect of time, 40
    concept of, 40
    conceptual critiques, 41
    contemporary acceptance of, 168
    evaluation, 76
    modern rational paradigm, 42, 76, 168
risk society, 80.1, 80.2. *See also* risk
Rohland, Eleonora, 38, 103

Sahel
    climatic shifts, 76
San Francisco earthquake of 1906, 54
sand drifts, 55, 56, 75, 83, 93
SARS
    and victimization, 138
Sartori, Giovanni, 71
scale of disasters, 25
    assessment difficulties, 29
    focus on material losses, 26, 27
    focus on mortality, 25, 27

Scheidel, Walter, 7, 70, 71, 154, 158
scope of disasters, 25
Second Pandemic, 69, 151
    heightened solidarity, 140
selective mortality, 129
sex and gender, 128
social profile, 125
    urban and rural environment, 126
Sen, Amartya, 13, 24, 88, 89
    entitlement theory, 25, 125
shocks, 2
    biophysical, 2
    concept of, 30
    effects of, 3
    human-made, 2
    nature of, 2
Silent Revolution, 92
Singleton, John, 31
Snow, John, 78, 114
    Broad Street pump, 114
societal collapse
    and moral decay, 142
    definition, 142
    environmental explanations, 144
    historiography, 145, 169
    transition and transformation, 145
Southern Africa
    inequality and droughts, 99
Southern Nyasaland, 4
    famine of 1949, 4, 88
Sri Lanka
    tsunami in 2004, 131
St. Vincent
    eruption of Mount Soufrière in 1812, 72
state coping capacity
    ideological underpinnings, 91
    regime type, 90
Steinberg, Ted, 97, 100
sub-Saharan Africa
    droughts, 55
    droughts and shifting blame, 140
    persistence of famine, 77, 78
    rainmaking rituals, 121

Tainter, Joseph, 142, 143
Tarr, Joel, 79
tragedy of the commons, 94

Ukraine
    Chernobyl nuclear disaster, 40, 167, 168
UNDRO (United Nations Disaster Relief Organization), 91

Index

Vaughan, Megan, 88
Voltaire
   and the Lisbon earthquake of 1755, 167
Vries, Jan de, 12, 178
vulnerability
   and economic commercialization, 83
   and economic diversification, 83
   and economic integration, 84
   and economic intensification, 82
   and human agency, 77, 163
   and inequality. *See* inequality
   and poverty, 96, 175
   and the powerless, 115
   assessment of, 66
   biophysical, 75, 76, 77
   connection to adaptation, 67
   determinants of, 24, 31
   development of the concept, 35
   indicators of, 67
   marriage patterns, 88
   pre-existing, 3
   technological lock-in, 115, 177

Wallerstein, Immanuel, 19, 165
White, Gilbert, 11
Wisner, Ben, 11, 81
World War II, 9
Worster, Donald, 12

Yersinia pestis pathogen. *See* Black Death

For EU product safety concerns, contact us at Calle de José Abascal, 56–1°, 28003 Madrid, Spain or eugpsr@cambridge.org.

www.ingramcontent.com/pod-product-compliance
Ingram Content Group UK Ltd.
Pitfield, Milton Keynes, MK11 3LW, UK
UKHW020306140625
459647UK00006B/65